The

Meaning of LIFE

If Life is a Journey You Need Good Directions

Thomas J. Miezejeski

Brookside Books

Toms River, New Jersey, 2002

Published by Brookside Books, 616 Brookside Drive, Toms River NJ 08753 732-573-0473

Library of Congress Control Number: 2002091996

ISBN: 0-972-0530-0-X

The author is grateful for permission to reprint the following copyrighted material:
Figure 3.1 Diagram of the Relatedness of Primates, page 21 from The Third Chimpanzee by Jared Diamond Copyright 1992 by Jared Dimond. Reprint by permission of Harper Collins Publishers, Inc.

Contents

Part II

Dedication

This book is dedicated to my parents Thomas and Julia and my daughters Noelle and Tara, who are the next generation. An important part of the meaning of life is simply passing along to the next generation, plus a little, what we received from the generation that gave us life. This is both our privilege and our responsibility.

Preface

What is the meaning of life? Does the human species have any special purpose in the universe? Does the universe have a purpose? We all think about these questions from time to time. Most importantly, we ask what is the meaning of my life. To ask why defines us as humans. No one has to encourage a child to ask, Why is the sky blue? or any of the other thousands of questions a child asks.

Some children grow up to become scientists and continue to search for answers. In the last few hundred years, science has discovered many things about the universe. We have learned the universe is very old and very large. Even to the nonscientist it is obvious the human body is very complex. Through science we have been able to learn that our bodies are far more complex than we could have ever imagined.

The intent of this book is to provide some of the information we have learned through science that can provide a new perspective on the meaning of life. Much of what we believe today about the meaning of life comes from myth and religion, which was set down in time before we had any alternative explanations. Gallileo was one of the first scientists to challenge the teachings of the Church and Darwin not only challenged the Church, but he also made us rethink our position within the entire spectrum of life on earth.

As science develops more sophisticated tools to study nature we are learning things that seem to be contrary to common sense. For example, Einstein's theory of relativity says that time is just another dimension of space, which is no different than the three dimensions that we experience in day to day life. Science has been able to observe phenomena, such as "action at a distance", which even Einstein refused to accept as possible. The concept of anti-matter is no longer just something in science fiction.

Although I do not make my living as a scientist I have had a life-long interest in science. However, you don't have to be interested in science to benefit from this book. Anyone who simply wants to be happy and feel they are living a meaningful life can benefit from the concepts presented in this book. Better yet, this book may clear-up some misunderstandings about reality that we have learned from myth or religion. The ideas from science are presented in simple language and they are presented with the assumption that the reader has no scientific training. On the other hand, people who like myself have a natural curiosity about science should enjoy seeing how information from many different branches of science are used to present a concise picture of what really *is*.

Some people might assume that this book takes an anti-religious position, but it does not. There is no question that harm has been done in the name of religion. The events of September 11, 2001 are just some of the more recent examples. However, religion, and myth before it, has done much to promote the social evolution of our species. The overall intention of religion is good, but most religions present their ideas and teachings as inflexible dogma that must be accepted on faith without question. Science has done much to improve the physical quality of our lives. In fact when a person is struggling to survive, they do not have much time to be concerned with such concepts as the meaning of life. However, most people we have managed to get past the point of mere survival. Thus, it is time to get past some of the ideas from religion that can stand in the way of a meaningful conscious experience. Our conscious experience, which we could also call our spiritual life, may not extend beyond this life. If this is true, we have even more incentive to maximize the quality of our current conscious experience.

Part two of this book looks at the personal meaning of life. It looks at what it takes to feel that we are living a meaningful life and how we use family and work to develop a sense of self worth. Ironically, the industrial revolution and the free enterprise system that we live and work in today is an outgrowth of science. Most of the increased comfort and security that we enjoy today comes from the practical application of what we have learned through science. In general, we have used science well to provide many goods and services to the average individual.

While this book hopes to encourage an understanding of science as a way of finding a meaningful life, it takes an opposing view to some of the ideas that our free enterprise system promotes as the sources of a meaningful life and a sense of personal self worth. Most people would assume that religion and material wealth are opposing sources of personal meaning. However, they have actually worked together to promote misconceptions about what is important to the meaning of life and happiness.

I have been fortunate to write a great deal of nonfiction during the last 15 years. In most cases the information that I passed along was more useful to the reader than to myself. As you might assume, writing this book has been a very rewarding personal experience, because I have been able to learn from the writing of many other people. I would like to thank them collectively for their thoughts and information, and I pass it along for your consideration.

How to Read this Book

According to a survey, a large percentage of people who bought Stephen Hawking's book *A Brief History of Time*, never finished reading the book. I think that while science can be wondrous at times, it is sometimes hard to understand, even when it is explained for the non scientist. This book relies heavily on what we have learned from science to provide a perspective on the meaning of life. If science couldn't provide a reasonably different perspective the Church would not have had such a problem with science when it first emerged. For this reason, I have devoted the front part of the book to higher meaning and the role of science in higher meaning.

If you find the front part of the book a little slow going please skip to the chapter on personal meaning and then try to move from there to the end of the book. I have tried to write each of the chapters so that they all stand on their own. Thus, I believe that much can be gotten from each by itself without having to refer back to other sections of the book.

Author's Note

Like most people, I have been thinking about the themes that are presented in this book for many years. Part of what makes us human is to think about our purpose and existence. However, we tend to take a casual approach to the topic especially when we are young and think we have many years to live. In a sense, we are so busy living our life that we take no time to really think about what we are doing with our life. We are almost on automatic pilot, the automatic pilot that was put in us while we were growing up, by our parents, our religious leaders, and our teachers.

About five years ago I encountered a personal crisis. I was faced with the need for cardiac bypass surgery without the health insurance to pay for it. Eventually I got the insurance coverage to pay for the operation, or else I most likely wouldn't be here today. I think that this experience is what most people refer to as a life changing experience. Naturally, this experience forced me to take a harder look at what really is important in life and what is the meaning and purpose of our existence.

For over 10 years, I have been making a living through writing, although it is quite different writing. I write market research reports on the telecommunications industry and I have also written a book on computer programming. About two years ago I started putting together some notes and accumulating materials with the idea that I might some day write this book. My experience with other writing projects, and what I have learned from talking with or reading about other authors made me wonder whether I could write anything that would be interesting to enough people to make the effort worthwhile.

Then on September 11, 2001 the whole world changed. I have been a New Yorker all my life and I have worked in and around the World Trade Center from a time when space occupied by the World Trade Center was a collection of stores selling electronics equipment. Thus, you could say I have a bias by attaching as much importance to the events that took place on September 11th. However, I believe that these events proved in a few hours that we live both in a global economy and a global village. We may have given lip service to the concept, but we also held back somewhat from these ideas because we didn't really live in a third world shanty town, and other people's suffering really couldn't reach us. We enjoyed the benefits of all the resources in the United States that have made our lives so comfortable and secure.

I present the ideas in this book to the reader as an individual, with the hope that it may help the reader take a different perspective on the meaning of life. After September 11th we are all now truly members of one human race that must be able to live with enough common purpose and meaning that we can survive as a species until we are overtaken by some natural calamity beyond our control.

Chapter One

Why Search for Meaning

"Without the gift of meaning we would never fully appreciate the gift of life. For that reason, if for no other, people should be encouraged to continue to ponder life's meaning. It is the question, not the answer, that is the real miracle. The quest for meaning alone enables us to be fully human."
Roy F. Baumeister

The Paradox of Meaning

What is the meaning of life? Do we have a purpose for being? Is there some all-controlling power that has set the universe in motion and has some ultimate purpose? As self aware or conscious beings we all ask these questions. Our capacity to ask these questions separates us from everything else in the world we know.

Although some say there are no absolute answers to these questions, we all need some limited and possibly inadequate answer to these questions in order to go on with our lives. Science, religion and philosophy are all efforts to provide some answers to these questions. Religion tries to provide us with some absolute answers while the answers from philosophy and science are more tentative.

Some philosophers, existentialists, say that science indicates there are no absolute answers.

The question about the meaning of life presents us with a paradox. There is little doubt that everyone wants their life to be meaningful, and at some level we all know what that means to us as individuals. On the other hand, when we think about the meaning of our individual lives as a whole, in no way considering the larger question of the purpose of life on earth, we have more difficulty understanding what that involves.

Humans have pondered the meaning of life since we were able to have conscious thoughts, but it appears that as we start the twenty first century understanding how we formulate a meaning of life for ourselves and as a species is of vital importance to the survival of our species. Due to our large brain size humans have come to dominate the rest of life on earth. Ironically our large brain size is most likely the reason why we can even ponder concepts such as the meaning of life. Thus, if you are a religious person you might say conscious though is both our blessing and our curse.

Some might suggest the question of the meaning of life is a problem for philosophers, or theologians, or some other experts can consider, my life is too busy to take the time to think about such issues. Many people appear uncomfortable talking about the meaning of life. When researchers engage people in open-ended discussion of the meaning of life it comes to a dead end fairly quickly for most people. It appears that the conversation is viewed as fruitless or we might learn something that could make us uncomfortable.

However, it appears that people are keenly interested in the meaning of life, if you look at the question from a different perspective. When you go into almost any book store the self-help section is the largest non-fiction section. In that section there are titles about happiness, how to be a more effective person or leader, strategies for success in a relationship or on the job, which are all aspects of living a meaningful life. General interest magazines and talk shows often deal with subjects relating to the meaning of life, although the specific topic might be about better relationships or careers.

In spite of all these books and other coverage, one must wonder if we are making any progress toward understanding the meaning of life. In a time of the greatest prosperity in our nations history, we have the highest divorce rate ever, problems with alcohol and drug addiction, and widespread use of antidepressant medications. It appears that people in government are more interested in maintaining certain political ideologies than they are in delivering services to the public, and it appears that widespread greed was a major factor in the recent Dot.com meltdown and other recent developments in business.

One must also wonder about our progress as a species when the United States, which as a country, has made more efforts to advance the rights and opportunities of its citizens and people around the world, is attacked the way it was on September 11, 2001. September 11th got our attention, but the problem has existed for some time.

With the exception of people who believe in reincarnation, we think we join the ranks of conscience humans as blank slates. Everything we know about the world we have learned from what other people have told us, what we have learned from the various communications media available to us to today, and from what we have observed ourselves firsthand.

We sometimes overlook or downplay the fact that we come into this world with basic instincts like all other animals. We are now learning that we carry a great deal of information in our genetic material that determines how our bodies grow, how we adapt to our environment and even what diseases could end our lives.

Thus, I hope to show how we acquire concepts about the meaning of life, and what we do to make our lives more meaningful, but I also aim to provide some information from science about the nature of the place where we live. This information has helped me come to some conclusions on my own about the meaning of life.

The objective of this book is not to judge nor is it to develop some unique theory on the higher meaning of life, but it does have a theme. The theme of the book is our species has struggled to find some meaning and purpose of life for thousands of years. We are making progress toward discovering what is or finding truth, which in turn is making life more meaningful for each and everyone of us and our species.

You should note that I have already used the term species several times, because I like to take a very broad perspective at times. As a result of what we have learned through science, one group of philosophers, existentialists, have come to the conclusion that life has no meaning on a higher level. We live our lives based on our beliefs about the way the world is. It may not make any difference in the life of an individual to know that the earth is not the center of the universe, but there are some things that we have learned from science that may prompt us to live our life differently, and more importantly, give more meaning to our life experience. Finally, I believe what we can learn through science will help establish higher meaning even if it only applies to our species. For most people that should be more than enough.

The Focus of This Book

If you were to do a search of the books that have been written on the meaning of life in recent years you will find that there are very few books written specifically on this topic. The lack of books on the topic is due to two factors. First, when science emerged using the scientific method, which tests theories by experiment, it was difficult to subject concepts relating to the meaning of life to proof by experiments. Second, the social sciences have examined concepts relating to the meaning of life over the last 100 years, such as feelings of self worth and purpose, but science has become so departmentalized that few have tried to take on the whole subject, even within the social sciences.

We can take information from many areas of study as input in establishing a purpose in life, and more importantly this information can provide a prospective with which to judge how meaningful our lives are or could be. Personally, I have found that an understanding of astronomy and genetics has been as helpful as religion, philosophy or psychology in understanding the meaning of life. Thus, I will draw on information from all these fields in this book.

At one time, it was easy for one person to take such a comprehensive review of knowledge, because individual fields of study were neither very wide nor very deep. Today, all information in one subset of a field of science, such as physics, is greater than the sum of all knowledge, less than 100 years ago. However, there are some common treads of knowledge and experience that run through all knowledge. For example, chaos theory is being used to explain things in economics as well as evolution. Also, we are learning some things through observation that make us rethink basic concepts in philosophy.

Today, it is almost impossible for someone to be a Renaissance man in the scientific community. In the days of Decarte, a person could be a philosopher, a physicist, and a mathematician all at the same time. In fact, Newton developed a new branch of mathematics, calculus, to calculate the exact values for the forces in physics that he proposed in his laws of motion. Now, it takes many years of study to be able to practice basic science in just one discipline. It also takes significant funding, and effort to add new knowledge to the existing body of knowledge. Given the limited life span of a person, no one can become a peer to his fellow scientists in more than one or two disciplines. Although he was an astronomer, the late Carl Segan was one man who was able to approach the status. of a Renaissance man.

In order to publish any "scholarly" work today in journals, such as Nature, the work must be subjected to peer review, which also discourages work that covers more than one discipline. Even if someone were able to develop new knowledge through the intersection of two or more disciplines there may not be another person qualified to critically evaluate it. I believe Daniel Dennett was able to make a significant contribution to philosophy with his book *Darwin's Dangerous Idea*, because he was has such a deep understanding of evolution and microbiology. Dennett also shows how on many occasions scientist's efforts to disprove concepts that seem to go against logical thought actually, in the end, support the concept they were trying to disprove.

As I mentioned earlier, the books published as self-help or popular psychology each tend to focus on one out of many aspects of the meaning of life. They also focus on providing quick results so they are presented in terms of lists or steps such as *The 7 Habits of Highly Effective People*, or the *Ten Laws of Life*. My intent is to help the average person gain a better understanding of what it takes to live a meaningful life, but I don't have any specific step by step program. Therefore, out of necessity, I have positioned this work between the scholars and the popular psychologists. I hope that I can give the reader the benefit of my eclectic review of many disciplines without misstating or misrepresenting any of the information.

The Meaning of Life Throughout History

Surveys show that people get meaning from many sources and that these multiple sources can lead individuals to have a confused sense of meaning. Some peoples' confused sense of meaning may also include some myths, that society has allowed to continue, because these myths support the purposes of society rather than the individual. In chapter two we look at the three sources of meaning; religion or myth, philosophy, and science. All three can be sources of higher meaning, which we discuss in detail in chapter three. Higher meaning is some meaning that comes from outside the individual, and is based on some higher authority such as God, logic, or nature.

Before science emerged about 400 years ago, religion and philosophy provided a code of conduct for how we should live in society. In addition, religion offered an explanation of any thing that was not within the power of humans. In modern society we have become so accustomed to living with technology and a basic knowledge of nature, that it is easy to forget that at one time we did not know what caused the change of the seasons or many other things in nature, that is commonly known by school children today. The things that are still not explained by science, such as the concepts of God, the soul and life after death, are still explained by religion, or to be more accurate, are accepted on faith.

When science first emerged, it presented a significant challenge to the authority of the Church as a social force in civil matters as well as spiritual matters. But science also challenged the Church as a source of all information. All study, not just matters of religion, were under the control of the Church. Besides providing meaning in matters of moral behavior, the Church also played a major role in defining a person's place in society, which is an important aspect of personal meaning.

Science eventually was a challenge to philosophy as well. Philosophy based its principles on what we could observe to be true. As the tools of science became more sophisticated, scientists were able to observe a part of nature that was beyond what could be observed by the unaided senses. Eventually science uncovered phenomena in nature that contradict the intuition we developed from unaided observation. For example, the concept of space-time, which says that time is another integral dimension of space, just like height, width, and depth.

Over the years, as scientific explanations have replaced religious explanations, some people have been reluctant to accept these explanations as true. Unfortunately, some explanations provided by science reduce our sense of meaning and self worth when these explanations are compared with what is provided by religion. Some people, including scientists, have been reluctant to accept these explanations and incorporate them into a personal meaning of life. Even today, some concepts such as life after death are beyond the range of science, and people rely on the religious teaching with regard to these concepts. It is easier to accept these ideas on faith rather than to confront the issue.

Finally, there are some concepts that we incorporate into our meaning of life, which are contrary to our personal experience, but which we allow society to trick us into believing are true. Acceptance of these ideas could be viewed as a form of collective denial. For example, studies have shown that parenting is a difficult task that usually results in reduced happiness for a couple. But society continues to encourage the view that parenting will result in increased happiness.

Personal Meaning

Since the social sciences play such a major role in personal meaning, which is the primary thrust of this book, chapter six is devoted to sources of personal meaning as well as the processes that an individual goes through to live a meaningful life.

According to the Dali Lama and a number of self-help gurus, the purpose of life is to seek happiness. Often when people question the meaning of life, they are asking why they aren't happy. Psychology is one of the first sciences to directly address some of the same issues that once were only addressed by religion and philosophy. Today, some people talk to their priest or rabbi to get some of the same counseling that other people get from a psychologist or social worker. Psychology addresses issues around our conscious experience as well as how we learn, which are some of the same issues that were addressed by the classical philosophers.

The Way Things Are - Law and Order

Approximately 400 years ago the concept of higher meaning started to undergo a transformation. Before the age of science we knew practically nothing about the laws of nature. We knew the heavenly bodies move through certain cycles giving us day and night and the phases of the moon. We had some ways to treat illness and injury, but many people died from all but minor injuries and sickness. Since we were able to engage in purposeful activities we assumed that the world we lived in was created for a purpose. In general we believed in the anthropomorphic principle, which involves projecting the way we behave to the way nature behaves or what controls nature, which some people call God, behaves. There was much that we did not understand, so we assumed that it was created by and under the control of something outside of the world we experienced.

Somehow, as time went by, some people claimed to have special knowledge about the world and how it works. Usually this information was obtained by some divine inspiration. Eventually, truth became true because someone said it was so. Even today, some people still rely on the Bible because it is believed by those people to be the word of God.

With the emergence of science, men had another source of truth. If something could be observed and tested under controlled conditions, in other words, repeatable with consistency, that information was also truth. Putting aside all the conflicts between religion and science, we now have a very large amount of information that we consider to be scientific truth, which we also call facts.

In this book, we are looking at the meaning of life in the physical world, since it is the only one we can experience. Therefore, I believe that the meaning of our life as a species on the planet is integrated with what we know about the order in the physical universe. Today, people have a choice of believing what was once taught by religion or believing what has been learned by science.

In chapter three we look at how people come to higher meaning. We also provide a brief overview of a concept developed by Daniel Dennett, which he has labeled universal acid. Dennett believes that Darwin's theory of evolution has demonstrated that it is possible to have **design without intelligence.** This concept may be the most important concept in philosophy of all time. Briefly, Dennett shows that something as complex as the human species could have developed without some pre-established design. We humans design an artifact to serve a purpose that we know beforehand. The irony of evolution is that even if it involves thousands or even millions of iterations it never has any foreknowledge of the next step. Evolution can only adapt to the conditions that exist in the current environment.

I believe that almost as important as the theory itself, is the fact that it was developed through observing nature. Thus, I believe that observing nature is an important aspect of developing both a higher and a personal meaning of life.

Although we believe what science tells us is true, we are often motivated by ideas about the natural world, which have more of their basis in old religious ideas than on what we know about nature. Thus, by providing some information about the universe as we know it at the beginning of the 21st century, you may gain a new perspective on our place in the universe and our purpose in life. This knowledge is also important to making the experience of living more rewarding. If we take a neutral stance toward events as they happen, we can live a life with less stress.

Our Place in the Universe

A part of understanding the meaning of life is understanding where we as a species fit into nature. Religion teaches us our place in nature and our relationship to some entity that it teaches is the creator. When science emerged it showed that religion was basically wrong with regard to what it taught about nature. From the church's perspective, the mere fact that it was wrong was a significant challenge to its authority. However, science has been also able to discover what is, or might be, based on observation. Religion provides a sense of security in a chaotic world. Science tells us about the world that we experience as we go through our lives.

In chapter four we look at the uncertainty and the apparent chaos in the universe. While some of the facts we have uncovered about uncertainty in nature may be unsettling to some people, I believe that knowing the laws of nature gives me a sense of where I stand. Our sense of meaning and self worth is integrated with our experience, especially with regard to how much control we have over our day to day experience. Religion and to a greater extent the economic system we live in, has taught us that much of our experience of life is the result of a system of reward and punishment for our actions. I suggest in this chapter that it may be beneficial to personal meaning as well as a guide to personal behavior to take into consideration how nature behaves. Basically, many of the good and bad things that happen to us are the result of random events.

In chapter five we look at the age and the size of the universe. In this chapter, we show how although the earth is a very small part of the universe we are all made of star stuff. While some people say that science cannot prove that life has no higher purpose, it does provide a sense of the beauty and complexity in nature. I also look at the possibility that life exists elsewhere in the universe. Some people seek to use science to prove that God exists and in turn provide more meaning to life. I believe that regardless of what we can prove through science, an increased knowledge of the universe makes life more meaningful. It appears that one of the things that makes us human is that we are curious. The fact that we know more about nature is a reward in itself regardless of any contribution that knowledge may make to our chances for survival.

Some people say that the universe was created for man or the human species. However, one must give a second thought to this idea if one realizes that the universe existed for about 10 billion years before our solar system was formed.

In chapter six we look at the kind of meaning that we think about first when we think about the meaning of life. A man once said "ask a man about the meaning of life and he will tell you about the meaning of his life". The concepts and information about personal meaning take up a large part of the space in this book. If the book was being written three hundred years ago, there would be no need to spend more than a few pages on personal meaning. Three hundred years ago, people, from the king to the peasant, knew their place in society and what their purpose was in life. It was already set on the day they were born. There were some exceptions, where a person would rise up a level or two, but they were truly exceptions.

Today, a great deal of attention is devoted to the self. The popular psychology section in the book store is also called the self-help section. Society and talk show guests tell people today that it is not only their right but their duty to find purpose in their lives. People are told to find out who they are inside, and them live a life that satisfies their inner needs, rather than some purpose that was given to them by their parents or learned from society.

These books and talk show guests are not telling people to be selfish. Instead they are telling us that we have to make ourselves happy before we can please or help anyone else. An important part of personal meaning is a person's sense of self-worth, and self worth must still be based on an ethical value system. People get meaning from many sources, but two of the major sources for meaning in everyone's life are work and family. These two sources are so important that we devote a separate chapter each to work and family.

Work is an important source of meaning, because we spend so much of our time in a work environment. In addition, for most people, work provides the major part of their identity. A person may spend almost as much time off the job pursuing an avocation, such as charity work, but people will most like tell you what they do to make a living when they are asked what they do. Women who work may spend more time caring for children and their household, but they still get an important part of their identity from work.

When we think about meaning as a thoughtful process, family is a major source of purpose in our lives. However, there is a more basic source of meaning, which comes from our sex drive. Even in a society that is becoming more open about sex, sex is still somewhat below the surface. Clothing, automobiles and many other products are sold on the basis of the sex appeal these products will provide their owner. The advertising never says it directly, but they put attractive people of the opposite sex in the ad. When people choose their work and place of work, the ability to be attractive to the opposite sex is important. People may not be looking to develop a relationship on the job, but they are still looking to develop relationships based on their job. At the very least, the right job provides the money to live an attractive singles lifestyle.

Some people may point to the use of sex appeal in advertising as an indication that we are living in a society that is too permissive about sex. I am not presenting this information to judge, but rather to highlight the inf m . fluenis basic drive has on the meaning of our lives.

There is a saying that "Life is a journey not a destination". As the subtitle on this book suggests, I believe life is much more a journey than a destination. Every day we are buffeted by so many forces, over which we have so little control, we have to put goals in perspective. More importantly, we may never achieve our goals. Our lives can be cut short by an accident or illness that prevents us from living, what most people would call a full life. Many of the self-help books explain to people, why in a life of abundance we suffer from a lack of fulfillment. They try to point out that we sometimes confuse fulfillment with happiness. Many people think that they will finally be happy when they get the right job, or get married to the right person, or they buy a house.

In my mind the common thread in many self help books is that life is a journey, and many of them actually say it in so many words. Not everyone has the same number of steps or the same number of points in their program, but there are some common threads that support the idea that life is a journey. In spite of the fact that we cannot control much in life, we should have a plan that is based on a reasonable evaluation of our capabilities and desires.

The reason that we are human, and the reason why we are at least one step above the rest of nature is that we have the ability to anticipate. All of the progress we have made as a species has come from our ability to anticipate. Therefore we should use this power to make our life more meaningful. An important component of a sense of meaning is a sense of being in control. Control doesn't always mean being able to change things. Control can involve avoiding the things that you cannot change and can hurt you, changing the things that you can change, and accepting things you cannot avoid and can't change.

Any book that examines personal meaning must look at how we should behave in relation to our fellow man. Although we experience personal meaning internally we still interact with other people. One of the original subjects of philosophy from ancient times has been morals and ethics. As discussed in the chapters on work and family, we get a great deal of meaning from our relationships with others.

As our ability to communicate and interact with more people over greater geographic distances the concepts that govern our interaction with people also change. Today, we talk about a global village, but several wars were fought in the 20th century that involved a major portion of the earth's population. The basic unit of political governance has expanded from the town to the region to the nation state. Today we recognize nation state as the basic unit of governance, but we are required to call on an organization such as the United Nations to address issues that extend beyond national boarders. The problems with the ozone layer and global warming are examples.

At the same time when we are experiencing forces that require a common purpose for the human species some people are also experiencing or exercising the **Final Freedom,** which is to decide for themselves what is moral and ethical. In the past, people set goals and relied on moral rules that were handed down from a supreme being or at least some higher authority than themselves. Today, we make moral decisions based on what other people require and what are the consequences of our actions.

The world is changing so rapidly that the old rules don't apply anymore and no one has had the time to establish new rules. Religion has lost much of its influence in secular society. The corporation, which took over some of the role of establishing meaning in life, is now loosing its influence to the global economy. Companies can no longer expect employee loyalty in return for life time employment.

In the chapter on individuals and society, I also reflect on what it means to live in a complex society and how that meaning changes as society becomes more complex and interrelated. An understanding of physical evolution gives us an understanding of the meaning of life, so too can an understanding of social evolution. I end the chapter with a description of an encounter of a relatively primitive society with the modern world.

Hopefully, by the time you reach the final chapter you will have already come to the conclusions, expressed in the final chapter. Some people may still come to other conclusions. What do you think?

Chapter Two

Sources of Meaning - Religion Philosophy and Science

People say that what we're seeking is a meaning for life. I don't think that's what we're seeking. I think that what we're seeking is an experience of being alive, so that our life experiences on the purely physical plane will have resonance's within our innermost being and reality , so that we actually feel the rapture of being alive.

-- from The Power of Myth, by Joseph Campbell

Myth and Religion

While researching this book, one of the most startling realizations I came to is, the average man has had access to information that would allow independent conclusions about the meaning of life, for only a very short time. Even today, most people still do not have free access to this information. One could argue, that we have only lived in a short period of time when people needed to be concerned with the quality of life. I agree, unfortunately, too many people are still concerned with mere survival.

We tend to think in terms of our current experience, but we are influenced by the thinking of the people who cared for us as we grew to adults, and they in turn were influenced by the people that raised them. Thus, the history of thought has an influence on our thinking, even when those thoughts are no longer a part of our most current philosophy.

This chapter examines how ideas with regard to the meaning of life have been established and communicated to us today. The first communication of such ideas was through myth. Although some myth is written down, the primary mode of communicating myth is verbal. For some primitive people today, the communication of myth has never moved beyond the verbal form.

When we think about religion today we tend to think about Christianity, Judaism, Islam or Buddhism. When we think about myth or mythology we think about stories about heroic adventures and quests. In fact, we often referred to something as a myth when it really didn't happen. But, for most of human history myth and religious parables were the only way that any information about the world around us was communicated. The Bible has been called "the greatest story ever told" In addition, through stories, parables, or myths one generation related to another generation what was good social behavior, which behavior was admirable or heroic, and which behavior was not acceptable.

As society became more sophisticated, myth became more formalized as religion. Each succeeding iteration borrows from the past and adds its own part. Most of our Christian holidays today have some connection to pagan feasts of the past. There is a connection between Christmas and pagan rites with regard to the winter solstice. There is also a connection between Easter and pagan rites of spring. It is also not a coincidence that major Jewish holidays also occur around these times.

Likewise, philosophy, while attempting to break from religious thought was influenced by it. In periods when religion had great political influence, philosophy was very much influenced by religion.

Finally, science as we know it today may not be perceived as an outgrowth of philosophy, but the first scientists were philosophers first. Non philosophers may think that philosophy is no longer relevant to science today, but as we will show in the next chapter philosophy can make a very practical contribution to science today.

Changing Relationship of Religion and Science

The primary theme of this book is: science can provide some insight into why we are here, the purpose of life; and most importantly, how we can make the time we have in this life most rewarding and meaningful. However, I do not imply that meaning can not be gotten from other sources. For example, myth draws upon the experience of countless generations to teach us about the meaning of life. One theme of this book is that experience rather than goals are important. Therefore, knowledge about experiencing life is also important.

Until about four hundred years ago, the only answers about the meaning of life came from religion or philosophy. Therefore, I believe that it would be helpful to take a chapter to discuss how religion, philosophy and science influence the way each person arrives at answers to questions such as the meaning of life.

From the very beginning of science in the 16th century there has been a tension between religion and science. This tension has been unavoidable, because on many occasions science has provided explanations for the workings of nature that have been at odds with dogma. Philosophy on the other hand, has enjoyed a more workable relationship with religion, since it has sometimes been used to support and in some cases enhance dogma. It is of interest to note that in recent times science is now also being used by some to enhance dogma. For example, some religions teach that man was created in Gods image, and that all God's creation was confined to the planet earth. If this were so, then there could be no life anywhere else in the universe. If there was life elsewhere it would raise questions about the nature of God and his relationship to mankind.

Today we know from science, there is a very high probability that life does exist in other places in the universe. Science has also shown that the laws of nature are consistent throughout the universe. Thus, while some information may contradict traditional teachings from the Bible, other new information can be used to support the idea that there must be some higher power that provides for the consistency in nature throughout the universe.

My intention is not to take sides in this debate, but rather to put the two sides into some context that will allow you to come to some conclusion on your own. Furthermore, I do not believe that it is necessary for an open-minded person to take sides in the sense that either religion or science can provide us with all the right answers. We can learn from both science and religion; however, some religious concepts can take us in the wrong direction, when thinking about the meaning of life.

I agree with some philosophers, and some scientists and even some people in religion, who say that we can not know with absolute certainty that the universe was designed and is under the control of some ultimate force, that is usually referred to as God. In other words, there are some questions that cannot be answered by science or religion. In the long run, we believe what we want to believe, and one of the most important points of this book is; we freely accept what we believe rather than having it imposed upon us as dogma.

As an outcome of people making their own decisions, some people have started to make a distinction between a religious God and a personal God. The concept of a personal God involves an individual's own ideas about the existence of a higher power and God's role in the universe. The concept of a personal God flows out of the increased emphasis on an individual's free thinking, and establishing their own personal meaning for their lives.

Religion Serves Several Functions for Society

Religion serves several functions for society. Religion provides an explanation for the unexplainable. By providing an explanation for the unexplainable we realize a sense of comfort. By accepting religion's explanations of the chaos in nature, we assume that we have some possibility of controlling nature by appealing to this power through prayer or sacrifice. Religion provides social control. Society uses religion through its laws to impose control on personal behavior. Finally, religion uses ritual to increases its authority, and provide a sense awe and community.

Religion relies on the power of persuasion to get people to adopt its teachings or concepts. Faith is an integral part of religion. People accept religious teachings on faith, because religion freely admits that what it teaches is beyond explanation, at least an explanation that we can understand. Because religion is not based in fact, it is possible to have more than one religion, each with different teachings and concepts.

Because religion cannot prove what it teaches, it must rely on a higher authority, which is God. Religious leaders, such as Moses, or prophets, or the Pope, assume the authority of God by claiming that they are relating information from God. Unfortunately for religion, human logic dictates that if some truth is from God it must always be correct and unchanging. This principle in turn leads to inflexibility in religion. If new information is available, that makes a teaching appear to be incorrect, religion has a problem with the change. If one teaching can be challenged by new information, the door is opened to challenge other teachings.

In the early years of science, the Roman Catholic Church was the dominant religion in western society, and the infallibility of the church applied to all matters, not just ethics and morals. One of the first scientists, Galileo, proposed that the Earth revolved around the Sun rather than the Sun revolving around the earth, which would be a natural conclusion from unaided observation. Galileo was subjected to a trial and subsequent house arrest for his proposal. It took the Catholic Church almost two hundred years to eventually remove his book, ***Dialogue***, that detailed his theory about the Earth going around the Sun, from its ***Index of Prohibited Books.***

In the nineteenth century, Charles Darwin was also the subject of great controversy for publishing his book ***On the Origin of the Species by Natural Selection***, which detailed his theory of evolution. By proposing that humans could have evolved from other species through a process of natural selection, Darwin was contradicting the Church's teachings, that man was created in Gods image, that man was unique among all animals, and that the rest of creation was created specifically for mankind's benefit.

As far as most people are concerned, Darwin's theory has been confirmed by what we have learned from fossil records and more recently from what we have learned from DNA, which provides the instructions for building all forms of life.

I present these two examples of the conflict between religion and science to show that science may be able to provide another perspective on our place in the universe. In turn, this new perspective may have some impact on how we live our lives. Until science was formalized, humans had no other source upon which to support order and social control.

As an aside, some religions over time have been more adept at dealing with change in spite of their inherent inflexibility. For example, the Catholic Church was able, eventually, to reconcile the discoveries of Galileo and Darwin with dogma.

Explanation for the Unexplainable

Until recently life on earth was very dangerous and unpredictable. We lived with the risk of starvation, death from storms and exposure to the elements, disease, and even death from other groups within our own species. Until the beginning of science we knew very little about how the world worked. Thus, we relied on myth and religion to provide an explanation of the world we live in and to protect us from adversity.

The human species is about 4 million years old, but our recorded history goes back less than 10,000 years. There doesn't seem to be any practical way to determine when humans developed the capacity to communicate through speech, but intuitively it should have been many thousands of years before we developed writing. The fossil record indicated that Homo sapiens, the last branch in evolution that lead to modern day man, emerged from Neanderthal man about 38,000 BC or about forty thousand years ago. One feature that distinguished Homo sapiens, us, from Neanderthal man, was our much larger brain size, which enables our capacity to communicate through speech. Speech and language not only enable communication, but they are also required for the complex though processes that take place in humans.

Information and Communication

Recorded history is very short and the period of widespread literacy is much shorter. During most of recorded history only a small percentage of the population had the capacity to read and write. The people who could read and write usually held some position of authority. It is also important to note that people in a position of authority usually were the authority on all matters religious and civil. Until very recently, any religious ideas as well as anything else, such as family or tribal history, was communicated verbally. Thus, all knowledge was passed from one generation to the next through the spoken word. Even today, parents still tell their children fairy tales, and much of what we learn about life in our early years is related to us through the spoken word. In terms of day to day activities, we become very proficient in speech before we start to learn to read and write.

It is very important to note how significant it was that verbal communication was the only form of communications. Before written communications or the ability of the average person to take advantage of information that was written down, people providing verbal communication had a monopoly on information. They controlled the information illiterate people could use in establishing any independent thoughts about the meaning of life and how we should conduct our life. People still have direct experience as a way of acquiring knowledge. Greek philosophy and some eastern religions today feel that this is a legitimate way of acquiring information, but, as we will show when we discuss philosophy, this technique has limitations, especially when dealing with such all-encompassing issues, such as the meaning of life.

According to one poll, the printing press was voted the most significant invention of the last millennium. Few people can argue with this selection. Without the wide spread availability of books, wide spread literacy wouldn't make any sense, or would be very hard to achieve. It is not surprising that the Bible has always been one of the most popular and available books, since the invention of the printing press.

The bible is often cited as the official source of religious authority, but an officially accepted version of the Christian Bible was not available until the sixth century. Until that time, Christian teachings and thoughts were contained in many books. Only some of these books were consolidated to create the Bible as we know it today.

At one time the Bible was the source of all knowledge. It described creation and provided a history of man. It set down the rules of life. It provides an answer to all the questions that a person can not answer himself. As we learn more about nature through science, we rely less on religion for answers to our questions, but most people still rely on religion for the hard questions. Most people no longer believe that the universe was created in seven days. However, science is still a long way from explaining consciousness, so most people still believe that humans have a soul that leaves the body at death. We are raised as children with a system of rewards and punishments. As adults, when we suffer from misfortune that does not appear to be the result of our actions, people ask for Gods help. When a loved one dies, it is comforting to believe that all souls will eventually be united in an after-life. In general, religion handles the hard questions. As we learn more from science the list of questions will get shorter, but there will always be some unanswered questions.

Social Control

We live in an environment where everything is either predator or prey. We live in an environment of survival of the fittest. Humans improved their abilities as predators by working together. Humans also reduce their chances of being prey by living in groups. Human offspring go through a long period of dependency on their parents, before they are adults who can live on their own. Our most basic purpose in life is to mate and pass on our genes to the next generation

All of these environmental and biological functions make us social beings. Humans have been able to dominate all other species because our large brains have given us the ability to communicate effectively with others, which also makes us very social animals.

Because we are social beings and because we are intelligent, we developed myth and religion for organization and to provide rules of interaction. Because we have intelligence, we need a rational explanation why we should follow these rules, and we need someone to provide us with these rules.

Western religions, almost universally, use scripture or some other documents as the source of the authority for the rules. This is a part of the process of setting the rules. In turn, the authority behind scripture is God, since most religions and myth teach that myth or scripture was revealed by God to the person relating it. Before philosophy and science, God was the source of all truth and knowledge. Since God was the source of all truth, this authority would naturally extend to an explanation of the world we live in even when it did not effect the way we interacted with each other as social beings.

There have been some common themes in myth and religion. Many we still have today.

- Most religions believe that there are one or more gods that control the physical world and they are superior and more powerful than humans. In other words, natural order and law come from these gods.
- These gods can be influenced by prayer or sacrifice to provide us with our physical needs, and protect us from natural calamities.
- There is a physical world and a spiritual world. People live for a time in the physical world and then their spirits go to another place, which is either better or worse than the physical world based on our deeds in the physical world.
- In some religions, where more than one life in the physical world is possible, the objective is to work to a higher state of being or even perfection.
- Man can control his actions and he is responsible to the group and to god for his behavior. Most laws governing behavior are negative. The rules are stated as thou shalt not...
- Things that we cannot control or understand are under the control of the gods.

Religion Can Divide as Well as Unite

Unfortunately, the inflexibility of religion can lead to conflicts between groups that hold different religious views, unless a society, such as the U.S. society, allows religious freedom. Almost all social behavior has some root in religious teachings. Thus, even in the U.S.

society, there are limits to people's religious freedoms. For example, most religions in the U.S. have a basis in Christian concepts; therefore any religious practices such as ritualistic sacrifices would not be permitted even on the principle of freedom of religion. On the other hand, since capital l puniment and killing an enemy in wartime are within the limits of Christian teachings, they are permitted in U.S. society.

Ritual

Religion also provides us with ritual. Besides ceremonies associated with prayer and sacrifice. There are ceremonies associates with the beginning of life and death. The are also ritual associated with marriage, attainment of puberty, and the preparation for hunting or warfare. The ritual provides importance to these occasions and sense of the community for those participating in the rituals. Ritual has been carried over into non-religious occasions. For example, students wear robes at graduations. The robes are one way of conferring honor on the students for their academic achievements. Judges wear robes in courtrooms. The judge wears robes not only to encourage respect for his position but also to encourage respect for proceedings taking place in the court. Two people could take marriage vows in private and the marriage would be just as valid, but a public ceremony confers a sense of importance to the occasion.

The main point with regard to rituals is that religion has become a part of our lives in many ways far beyond dogma and social control. Later in the book we will talk about the meaning and sense of self worth that we get from being a part of a family and the relationships that are a part of family life. While we may be able to establish some new ideas about the meaning of life from knowledge gained from non-religious sources, religion will always make a contribution to a meaningful life.

Before science, for all practical purposes, it was meaningless to think about the meaning of life, or our goals in life, or how we should live our lives outside of what we were told was required by religion. Because living was so hard, many religions taught that the primary purpose of this life was to prepare for a better life after death.

Practically all of our moral and social codes flow from religion and we must follow those codes if we want to be a part of society. Some people incorrectly believe that if we conclude that God doesn't exist or if there is no life after death that we are also not bound by moral and social codes that have been based on religious teachings. Regardless of what we may ultimately learn from science, we will most likely still require the ritual associated with religion, although it may be secularized, since it is so ingrained in our day to day life. Today we enjoy the benefits of technology, which make our lives less risky, but we still live in a chaotic world. We still need the ritual to provide some sense of order and security.

The United States was founded on the principles of separation of church and state and religious freedom, but the United States from its founding to this very day is strongly influenced by religion. Polls show that more people in the United States, from 92% to 97%, believe in God, which is a higher percent than in almost any other country in the world. In some countries, the belief in God is only in the 80% range. This appears ironic, because the U.S. is the country that has benefited the most from the advancement of science.

Unfortunately, many people still live with the all or nothing approach to religious thought. In addition, other people, who consider themselves to be more open minded, may still be influenced by concepts that are so a part of religious thought that they do not realized that they have these biases. For example, a poll of scientists shows that 60% of scientists still believe in God, although in many cases they believe in a personal God rather a religious God. These scientists feel a need for religious concepts either for personal security reasons or for social convention.

Stated as simply as possible, religion has and continues to have a very strong influence on our thinking about many things relating to the meaning of like. Although religion may no longer hold the central position in our society it still has the power to influence our thinking. Although religion may provide some sense of security with regard to the chaotic world we live in, it may also be holding us back from an understanding of what ***is***. An understanding of what ***is*** can result in a happier and more meaningful life. I believe that religious thought can be confining with regard to human capabilities, whereas an understanding of what is from science can be liberating and awe inspiring. Science can provide us with an awe of nature that is far greater than the awe that can be provided by religion.

Meanwhile, we must retain the social benefits that we enjoy from religion either directly or indirectly. We will most likely require religion or something like religion, because we are conscious and social beings.

Philosophy

While philosophy approaches questions such as the meaning of life, and how we should lead our lives from a different direction, many of the fundamental questions in philosophy are the same. While philosophers have examined such issues as the existence of God and the free will of man, there has been less direct conflict between religious teaching and philosophy. However, in the twentieth century some conflicts between philosophy and science have also developed. These conflicts have not involved science proving philosophy wrong, but rather some things we have learned in science are difficult to reconcile with basic philosophical principals.

While a simple definition of philosophy could be that philosophy is a study of truth, a more detailed explanation is difficult to compose. Hobbes defined it as “a knowledge of effects from their causes and of causes from their effects”. Hegel defined it as “the investigation of things through thought and contemplation". Hegel makes an important distinction between science and philosophy. According to Hegel, philosophers do their work through thought and scientists do their work by direct observation.

At first this may seem to be a simple enough definition, but on further examination philosophers get into many problems of defining how we make observations and how we know through thought. In fact, many of the classical philosophers were scientists as well as philosophers, which would imply that it is difficult to make a clear distinction between science and philosophy. Even today, the study of physics, cs, whiis seeking a theory to explain all the forces of nature, and astronomy, which is exploring the big bang, are only a few steps removed from philosophy's study of what is the ultimate cause and source of everything in the universe.

History of Philosophy

The Greek philosophers studied the fundamental structure of the world and the principles governing the order of events. They developed mathematics and some basic principles of physics and astronomy. However, the primary focus of the Greek philosophy was to explore such problems as the aim and meaning of life, the ethical principles of conduct, and the best ways to organize society and provide for government.

During Medieval times Christianity took over a large part of the world and it introduced a new moral code and its own ideas on the meaning of life. During this period, the Church professed to have infallible answers to such basic questions as the meaning of life and how men should conduct their lives. During this period, philosophy was used primarily to support the principles and teachings of the Church rather than to provide for free and open exploration through independent thought. Philosophers concerned themselves with questions such as; if God knew the course of all events, were man's action predetermined. Or how could men have free will if God knew the course of all action. These issues would not have been that significant if the Church was not so intent on holding men accountable for their actions. These questions were essential to issues such as guilt and evil.

Philosophy experienced a gradual revival in the 16th and 17th centuries as the theory of Copernicus and the observations of Galileo, through the telescope he invented, confirmed that the sun, not the earth, was the center of our solar system. This period marked the beginning of science and a break of philosophy and science from the control and dictates of the Catholic Church.

This break from the Catholic Church was not without a cost to the average man or the intellectual. The problems that started then still plague us today. The Church exercised control over society, but it also taught that mankind would be rewarded with eternal life in heaven for all the sacrifices and suffering that we endure in this world. It taught that a man's soul, which was created in the image of God, would go to heaven. Because man had a soul he was superior to all material things on earth. Challenging the teachings about how the universe worked created some doubt as to the existence of God and a reward in a life after death, and at least to some extent diminished mankind's stature in the universe.

Although in later chapters we do not try to prove the existence of God through the use of science, we do try to provide a perspective on the meaning of life in the here and now. It is hard to imagine what life would be like after death, if it has no physical reference. Furthermore, science has shown that time is just another dimension of our physical world. Thus, just as many philosophers and scientist believe that it is impossible to prove or disprove that God exists, it is equally impossible to prove that there is a part of each of us, which we call the soul, that will carry on after our physical death.

Avoid Conflict with Science and Religion

Philosophy freed from the constraints of medieval times found itself with a new special task. Just as the task of medieval philosophy had been to reconcile conflict between philosophy and religion, the task of Renaissance philosophy was to avoid conflict between science and religion. For example, while science was bringing into question the ability of men to have free will, Decarte was looking for things to maintain the freedom of the human will. As a result of working on this problem, it became obvious that if philosophy was to come to different conclusions than science, it must employ different methods. Furthermore, if philosophy was to reach different conclusions than science, philosophy must be able to prove that its methods were more reliable than science. It should be noted that in Decarte's time, there were none of the instruments, that science uses today to aid the senses. This was a critical problem, because the unaided senses can lead to erroneous conclusions.

If knowledge is not handed down to us from a higher source, philosophy must be concerned with questions such as, are their different kinds of knowledge, and how do we arrive at that knowledge. For example, Plato believed that certain information was inborn in our minds in some form of memory that it simply needed to be recalled. Other knowledge could be innate in the mind at birth and then formed as a piece of marble is formed into a statue. It seems ironic that although these concepts were eventually dismissed, science has recently discovered, that all living things contain DNA, which is in fact a huge amount of stored information that is the blueprint and instruction manual for all living things. Science today with sophisticated instruments is now able to decode this information. In addition, we are just now learning through science that all humans have the capacity for language at birth. When we learn to speak, we in fact are using an inborn capability. At birth, all human babies have the capacity to learn any language, but soon after we start to learn a specific language, we lose our facility to learn any other language in the same way that we learn our first "native " language.

The Source of Knowledge

Empiricists believed that knowledge only comes from experience, which is precisely the method of science. At this point it, would be helpful to review what most people know about the world today, but was only anticipated at the beginning of science. We live our life in a human size world. The objects that we can experience with our unaided senses exist about in the middle of two other worlds that can only be experienced with the aid of instruments. The two worlds are: 1. The world of atoms and molecules, which is millions of times smaller than the smallest thing that we can experience directly with our senses, both in terms of size and intervals of time. 2. The world of stars and galaxies, which is millions if not billions of times larger than we can experienced directly, also, both in terms of size and of intervals of time.

We will discuss the age of the universe is another chapter, but it should be simply said that 15 billion years is something that can only be perceived as a concept. While all objects in the universe are governed by the same laws of physics, one aspect of these laws applies to very small objects, another to human size objects, and another to the world on an astronomical scale.

Since philosophy can only arrive at knowledge from experience in the human size world with human size intervals, certain laws that may be defined as a priori or intuitive in the man size world can be proved to not be a priori, in either of the other two worlds that exist on either side of the man size world. For example, in the man size world we think we see cause and effect running through everything, because the phenomena of the man size world seem to conform to the law of causality. When in reality all we perceive are things that obey statistical laws that give us the impression of cause and effect. We could not imagine anything else because we cannot experience anything else.

In summary, philosophy learned that questioning our own minds through philosophy will only tell us truths about how our own minds work. Seeking knowledge through observing cause and effect will only tells us about perceived cause and effect, if we do not have instruments to assist our observation of true cause and effect. In order to discover the truths of nature, the pattern of events in the universe that we live in, the only sound method is to go out into the world and question nature directly, which is the system that has been proved many times when it is used by science. The recognition of this fact has actually brought philosophy to a closer relationship with science.

Science and Philosophy Reconciled

Reflecting on this fact, Sir James Jeans, in his book, *Physics and Philosophy*, which analyzes the relationship of Physics and Philosophy says "This may seem to suggest that philosophy should have not only the same methods but also have the same aims and, broadly speaking, the same field of work as science. But the distinction mentioned at the beginning still holds true. The tools of science are observation and experiment; the tools of philosophy are discussion and contemplation. It is still for science to discover the pattern of events, and for philosophy to try to interpret it when found."

We might add that to large degree that is the purpose of this book. We believe that the knowledge of what science tells us about certain aspects of nature can in turn help us come to a personal understanding of the meaning of the facts to our lives.

The material presented here provides only a minimal amount of information presented by Sir James Jeans to show the different approaches used by science and philosophy to arrive at truth and to put science in an overall context of the way that we as humans arrive at truth. I suggest that if you are interested in a more extensive and rigorous discussion of the difference between the approach and methods used by philosophy and science that you read Physics and Philosophy by Sir James Jeans. His book was originally published in 1943 and republished in 1981. The fact that the book is still relevant after almost 60 years of advances in science is an important endorsement of the validity of the observations that he makes with regard to philosophy and science.

Science

Before science, men often ascribed human characteristics to inanimate objects. They believed that the moon and the stars through their movement could have an impact on men, or that they had personalities, like the real people that they associated with. These personalities made them reward or punish mere mortals according to the favor they could garner with them. The key point was that men believed that they were at the mercy of whimsical nature.

Scientific Method

Science is an effort to determine the pattern of events as they govern or occur in nature. In their very first exposure to science most students are introduced to the scientific method, which has four steps:

1. Observation and description of a phenomenon, or group of phenomena.

2. Formulation of a hypothesis to explain the phenomena. In physics, the hypothesis often takes the form of a causal mechanism or a mathematical relation.

3. Use of the hypothesis to predict the existence of other phenomena, or to predict quantitatively the results of new observations.

4. Performance of experimental tests of the predictions by several independent experimenters and properly performed experiments

The underlying principle of science is the assumption that there is uniformity in nature. Given that the circumstances are the same, what we have experienced once will be the same the next time that circumstances are the same. Once man started to see this uniformity they realized that understanding this uniformity could make their daily lives safer or more comfortable.

While formal science started about 400 years ago, one could say that science started with the Homo sapiens. Archeologists have studied the tools used by Neanderthal and early Homo sapiens and they have found that although Neanderthals had basic tools for about a million years there was little improvement in these tools. Tools produced by early Homo sapiens showed improvement over Neanderthal man and over a period of less than 100,000 years Homo sapiens now use computers, telephones, and heavy machinery as tools. Almost all the progress toward better and longer lives made by humans, since humans branched off in evolution, is due to science. Humans today are not very much different than the first human from a physiological point of view, including brain size.

Tools of Science

Most historians agree that formal science started with Copernicus, and Galileo. The important contribution made by Galileo was the telescope -- one of the first tools used in science. The telescope allowed Galileo to enhance his senses, namely eyesight, to make more detailed observations of the sun, moon and planets. Galileo was able to prove Copernicus's theory that the apparent motion of the sun, moon and stars across the sky resulted from the daily rotation of the earth, while the motion of the planets through the stars could be explained by their revolution around a fixed sun. Galileo was even able to determine that the seasons were the result of the earth presenting a different part of its titled axis to the sun as it revolved around the sun by charting the apparent movement of sun spots over the face of the sun.

We each live a mental life in a house we call our body. We receive communication from the outside world through our sense organs eyes, ears etc. Through these senses we get knowledge of the outside world. Thus, the whole content of a man's mind can consist of three parts, a part that was in his mind at birth, the part that has entered through the senses, and the part which has been developed by the process of reflection. As we have discussed earlier in this chapter, philosophy has not been able to show that the first part exists at all.

Whenever a person increases the content of his mind he gains new knowledge. This occurs each time a new relation is established between the worlds on the two sides of the sense organs, the world of ideas in a individual mind and the world of objects outside individual minds, which is common to us all.

The study of science gives us new knowledge. For example, the study of physics gives us exact knowledge because it gives us exact measurements. Physics tells us the wavelength of various parts of light or the densities of various materials such as water, air, or iron. These statements import real knowledge into our minds, since each provides a specific number, the concept of which is already inside our minds, with the value of a ratio which has an existence in the world outside; this idea of a ratio is something with which our minds are familiar. Thus, the statements tell us something new in a language that we understand.

Any science uses an accumulation of facts as its raw material. Science then organizes these facts, and sometimes reduces some of these facts to general rules or formulas. Since we assume that nature is rational, by organizing the facts we can show certain features. In order to communicate our findings or ideas about these facts we must have a language of terms and ideas with which other people understand. In the same way, if an interpretation of the workings of nature are to mean anything to us, it must be in terms that are already in our mind.

In order to facilitate our understanding of facts about a particular subject we often organize this information into a model. In some cases these models take on a geometric representation, in other cases they take on a mechanical representation. For example, we may represent electrons and neutrons on an atom as a miniature solar system. These models help our understanding of the phenomena, but may in no way be actually related to the reality of what exists. As science has explored the increased complexity of nature it now often must confine itself to mathematical descriptions of nature, since our experience of the physical word has no analogy to the phenomena studied. For example, as humans we can experience only three physical dimensions of space.

The Progress of Science

Science usually moves in small steps. We learn new information that fits in with other information that we have learned previously, or new information supports some concept that we would expect to logically follow other information already known or confirm a theory that has not been measured or experienced with the aid of instruments. It is important to note that what we can learn through science is becoming increasingly dependent on the tools that we can develop to either increase the sensitivity of our perceptions or to help us process the data that we gather through experiment and observation. The electron microscope would be an example of a tool in the small world, and the Hubel telescope on the level of galaxies. Observations made through the Hubel telescope have allowed scientists to confirm or disprove many theories about the origin and development of the universe that could only remain theories until hard evidence was gathered to prove or disprove them.

Occasionally, science experiences major breakthroughs that, at first, seem to be untrue, because they are at odds with what we have been able to learn up to that point. Although the term has been overused in recent times, these changes are referred to as paradigm shifts. These are changes, which shift our whole form of reference, with which we organize or process new information. Moving from the geocentric system to the Copernican system of astronomy could be considered the first of these shifts, which in this case, established science as a form of study. The second major shift is the Darwinian theory of evolution, which said that our human bodies were an adaptation of the bodies of other living things, rather than something specifically created for humans. While these paradigm shifts are significant they are still related to the world that we know on the man size level. However, recent discoveries such as Relativity, which tell us that time is just another dimension of the time-space continuum or quantum mechanics, which tells us that light is, at the same time, both matter and energy. Perhaps we will experience another paradigm shift in the future, which will make everything comprehensible to our human minds.

We have learned so much in the last 100 years that at times some people believed that science would run out of things to study. Then we come to a deeper understanding of nature, and people wonder whether we will ever be able to understand everything. These shifts in views have lead some to ask "Is the world material or mental in its ultimate nature? Is the world we perceive in space and time the world of ultimate reality, or is it only a curtain veiling a deeper reality beyond". Mathematics and physics are now considering theories that would require as many as eight or ten dimensions of reality. On the other hand, some scientists suggest that reality cannot be described by some all-inclusive fixed formulas, but only as the workings of many small algorithms that interact in almost infinite complexity.

While science may have provided an explanation of reality that is at odds with what religion once told us, it appears that science can only go so far to explain everything. However, we believe that it has done an acceptable job of explaining reality at the man size level. Some people believe that we may someday have an explanation for everything, while others think that our very nature is such that we will always have something to study. There is a debate in science whether all the laws of nature were established at the time of the beginning of the universe, which may the time of the big bang, or whether new features of nature are evolving the way life evolves. In either case, we do know some things today that can help us live our lives today. We are a part of nature and we live in nature. Therefore a study of nature should tell us some things about the meaning of this life. Thus, we will present a meaning of life from the perspective of science today in the chapters that follow.

Science and Religion

Science got off to a bad start as it relates to religion. Science has provided alternative explanations to things that once were only explained by religion and then accepted by faith, because they really weren't explained in terms that man could relate to his every day experience. Christian fundamentalists feel that science and religion are mutually exclusive, at least on some topics such as evolution. You have to accept either that God create the universe the way it is described in the Bible or accept the fact that humans have developed over a long a a long pere through a process of natural selection. A few people even believe that science and the things we develop through science are inherently evil, or they present such an obstacle to saving your soul, that we must live without the benefits most people enjoy from technology.

Other people have worked in science and have tried to use what we have learned in science to prove the existence of God. In a sense they are trying to come to some reconciliation between religion and science, because science presents some issues that are difficult to accept physiologically. Science presents the possibility that there is no life after death and that man was not created to dominate the universe. For example, scientists have determined the conditions that were present just before the big bang occurred. They have determined that in order for the universe to go on expanding, after the very first few seconds of the big bang, various opposing forces had to be so balanced, that it is difficult to believe the big bang could not have been possible without the plan and design of a higher power such as God.

In 1950, Pope Pius XII established that the Catholic Church would no longer take a position on the findings of science, and he limited its authority to spiritual matters. Unfortunately, it is not that easy. The Church still takes a position on what is permitted and what is not permitted in the area of human reproduction. In the latest debate about the use of cloning or genetic engineering, people have declared that we should not be doing such things because men would be playing God. It's just not that easy to separate science from religion.

Recognizing that it will be difficult, I still ask that you read the remaining part of this book with an open mind. Try to think about the points raised not as an alternative to something that you may have already learned from religion but as some new information that makes this life on the planet earth richer and more rewarding.

At the beginning of formal science, scientists risked their life to present what they believed was the truth they had learned about nature. Today, scientists are rewarded for doing good science as long as they are working in areas that appear to provide good without bringing basic religion values into question. Religion can also be used as a way of creating a strong coalition for a common cause. The most universal common cause is our continued existence on the planet both individually and as a species. However, when America was attacked on September 11, 2001, the words "God

Bless America" could be seen posted on a wide range of churches, businesses and even public facilities. In reality it was a rallying cry of us against them in the war on terror.

Although I am not presenting any of this information as an opposing view, the information may be troubling. Regardless of our religious belief, our mortality is a difficult thing to face. When we are young we don't even think about it, and when we are old its comforting to believe that we some how carry on after death. Although it is our Constitutional right to believe anything we want it is also our human constitutional inclination to reject anything that may in any way question our system of beliefs or our sense of security.

In this book we will look at some things we have learned from science which may enable you to establish your own position on some of these issues. However, in some cases we show that society may have to set some guidelines or controls on what we can do through science, because the actions of individuals will have wide reaching impact on the human species. Our species is the beneficiary of the evolution of all life on this planet. We did not ask for this position, but now that we find ourselves in this role, we are responsible to do our best to continue our species and all life on earth. In the time scale of all life each generation only has control for a microsecond of the total process.

Science and Life

In western society we live with science all day every day. When just one of the modern tools that we use such as electricity, communications or the automobile is taken away our lives becomes more difficult. In fact, our lives as a society would be impossible without technology. Without the increase in productivity that we enjoy in food production as a result of science we would all starve. One hundred years ago most people worked on farms and there was only enough left over after feeding the people on farms to support a small percentage of the population in cities. Today only a very small portion of the population is employed in the production of food. Most people are employed in providing other goods and services.

If we have used science and technology to provide for our physical needs why shouldn't we allow what we have learned about nature help us live more rewarding lives in terms of understanding our purpose in life.

Non Religious Objections to Science

Some people object to science not because it questions anything that they may believe as part of some religious doctrine, but because the effort to explain everything in terms of particles, forces, or the successive change of life as an adaptation to environment is dehumanizing. As humans, we enjoy music, art, humor and many other things that could never be explained in terms of particles etc. Scientists may try to respond to these objections by pointing out that science reveals the beauty in the complexity and design of nature, but I don't think that will satisfy too many people that believe that we have a soul that is separate and apart from our physical bodies. However, I think that most of these people will still admit that we still live day to day in a physical world that sometime makes us feel either physically comfortable or at risk. The physical world does much to make us feel happy or sad. My hope is that understanding how the world works may in some way make our lives more meaningful even if it cannot lessen the burden that we must bear at times.

Chapter Three

Higher Meaning

Meaning We All Live With

What limit can be put to this power (natural selection), acting during long ages and rigidly scrutinizing the whole constitution. structure, and habits of each creature, -- favoring the good and rejecting the bad? I can see no limit to this power, in slowly and beautifully adapting each form to the most complex relations of life.

-- Charles Darwin

Higher Meaning & Personal Meaning

When people talk about the meaning of life, they can either be talking about the meaning of life to them, in other words personal meaning, or the higher meaning of life on Earth. Now that science tells us that life most likely exists somewhere in the universe, we have to think of the higher meaning of life in even a larger context. This chapter deals with the higher meaning of life and chapter six deals with the process of establishing meaning and self worth on a personal basis. Regardless of the source of our higher meaning it has an impact, either consciously or subconsciously, on the process of setting personal meaning.

Purpose and Meaning

This books addresses the meaning of life, but before we get into the details of the topic it would be helpful to define the words meaning and life, because both words can mean different things or can be used in different contexts. Since a synonym for definition is meaning we are at risk of confusion from the very start.

There are three ways of defining the sense of the word meaning. A synonym for one sense of the word is purpose. Purpose in turn implies some goal for an action, or use, if we are referring to an object. It is this sense of the word that most people think about when they ask what is the meaning of life?

People would be asking the same question if they said why am I here, or why does the universe exist. In other words, what are my goals, if we are talking about the individual, and what is the end objective of the universe, if we are talking about a more comprehensive meaning of life. This sense of the word also implies that there is a causal connection between events. When science studies the universe it is also looking for meaning in the universe.

The second sense of the word is synonymous with intention. We say that a person means well. Or we may ask what did he or she mean by that. What this means is that people express their intentions in words or actions.

The third sense is what I am doing right now. I am establishing some order or I am transmitting some information about different words.

With regard to life there are two major senses of the word. In the broadest sense, life implies a state of being where the thing that is alive or has life in it, is changing. Plants react to the environment and do all the things that we associate with growth and reproduction. Animals do all the things that plants do with the added capability of mobility. The human species evolved out of other living things and interacts with other living things so man is a part of life in this sense of the word.

A second meaning of the word life is the human experience on an individual basis or our interaction with everything else that makes up the universe. While this experience is going on we are alive. When the experience stops or is interrupted we are either unconscious or dead. Some people refer to life after death, which is the subject of many religions. For the purpose of this book I am referring to our conscious experience only.

Amoll the living things on earth the human species is the only species that has the capacity to consider such concepts as meaning and purpose. We are the only species that can look at things around us and can make connections between events and different parts of our environment. Humans engage in purposeful activities, such as raising crops and building structures. As a species we continually learn about the basic forces of nature, and even about how we interact with one another in society. Later in this chapter we explore the possibility that evolution may not follow any design. However, it is quite obvious that humans have the capacity, as a result of their intelligence to plan for the future and engage in activities that have a purpose. Ironically, evolution may not have a purpose, but the process of evolution has enabled purposeful activity. We continually apply this information to create **artifacts and memes** that improve our day to day life and increase our chances of survival against the forces of nature.

Other species engage in activities which could be viewed as purposeful, such as birds building nests and spiders building webs, but we attribute these activities to instinct. In other words these species are not aware that they are engaged in the activities to active an end objective, such as a place to raise young or to ensnare other insects.

As children, we first learn what things are and then build upon this by learning how things are used and how they interact. We gain much of this information without even having to ask. Although children will eventually ask such age old questions such as "Why is the sky blue?" Although we are not specifically taught that everything has a purpose, we eventually come to believe that everything does have a purpose. In fact, when adults start engaging in activities that do not appear to have any purpose we assume that they have lost touch with reality, or in some way are irrational. While some people may be able to get by day to day with on level of meaning, they often refer to some higher level of meaning when they experience something that does not, on the surface, appear to have a purpose. The death of a child or the death of an adult in the prime of their life are such events that prompt people to look for higher meaning.

As we will show in the next chapter, we live in an uncertain world and we live in a world that is constantly changing. While the world is uncertain and changing, we also seek to gain security from a sense of permanence that we expect to come from higher meaning. We take comfort in the belief the universe is an orderly place and that all things happen according to some plan, although we may not be aware of the details of why they happen.

As individuals and a species we have come to learn that the world is a very complex place. Over time, we have observed some things that occur with regularity and we conclude that there is some causal connection between events. If we bang one stone against another stone, there is a good chance that one of the stones will break into smaller pieces. An understanding of the outcome of various events increases our chances of survival in a world where there is much uncertainty.

However, we also learn that there are many things that we cannot explain or understand. In fact we have learned that some things may be unknowable. For example, Heisenberg's uncertainty principle states that some things in physics are intrinsically unknowable. As we will discuss in the chapter on uncertainty, in some cases, the best we can do is determine the probability a particular event will occur, or we cannot identify which particle will be the actual particle to change its physical state.

Throughout our history of discovery we have learned that we are able to explain more and more in terms of cause and effect. Ironically, the more we learn the more we realize the world is more complex than we could have even imagined.

Religion

Since the beginning of civilization, Religion has provided an explanation of the unexplained even if it simply told us that some higher power than man or nature created and controlled the things we did not understand. In accepting religion, people accept there limitations and accept on faith any explanation provided by religion. Thus, religion was the first source of higher meaning for humans and it continues to be the a source of higher meaning for many people regardless of how actively they take part in the practices of some organized religion.

Until the last few hundred years, life has been viewed as various levels of suffering, especially for the average person. Most religions teach that there is some existence after death and that at the very least that existence after death will be free from the suffering we know in life. Although all religions do not teach that our happiness in an existence after death is a reward or punishment for our actions in this life, many religions do have such teachings. Thus, these religions have used the belief in reward and punishment for our actions as a means of social control.

The Role of the Self

In the United States we live under a system of government as well as a social philosophy that places a great deal of emphasis on the rights and the autonomy of the individual. We as individuals cede to the government only those powers that we feel are absolutely necessary for an orderly and efficient society. Most political scientists credit this emphasis on the individual for our prosperity. When people are able to decide for themselves what they want to do with their lives, they will work harder not only because they benefit from their efforts, but also because they feel that they are in control of their destiny.

Before the industrial revolution, where people were born basically determined how they would live out their lives. If a person was born on a farm, as children they were expected to work on the farm. In other words they had to make a contribution to their family unit. If a boy's father was in a trade he was expected to follow in his father's footsteps. It is important to remember that until recently white males were the only people that had any chance at self determination. Women and slaves were viewed more like property than people when talking about self determination. In any case, the concept of self with regard to a higher meaning in life was much less important. The meaning of life was much more integrated with the meaning of a family, and the church. In fact, people who thought too much in terms of the self could be guilty of sins such as pride and greed.

Existentialism

As science has learned more about how life has developed on earth and how the universe itself has evolved many scientists have recognized the limitations of what science is able to tell us. Some people have come to the conclusion that science tells us that their is no higher meaning to existence. The universe is nothing more than the result of random events with no higher purpose. Philosophers that hold this belief call themselves existentialists. While some people would say the extreme existentialists believe that life is pointless, most existentialists believe that people live with meaning in their life. The key point of existentialism is -- meaning does not come from some external source of higher meaning such as God, the government or genetic destiny. Existentialists believe that everyone sets their own values and meaning to live their life.

In the study of the origin of the universe, science has been able to determine what happened in terms of physics up to a very small fraction of a second before the big bang. Many scientists believe that they will eventually be able to determine what happened right up to the instant of the big bang. However, those scientists along with most other scientists believe that it is not physically possible for us to know what came before the big bang, if in fact anything did. The deeper and more significant question, which we will not learn is whether the hand of God involved in the creation of the universe.

When religion played a central role in people's everyday life, religion took upon itself the role of explaining all things. Today, we live in a world where some things are considered part of secular society or outside of the church. Therefore, we are comfortable with the explanation of many things from science or more importantly outside of the religious domain. Today, most people believe that physical illness should be treated by science. In other words, they believe illness should be treated by physicians with pharmaceuticals or medical procedures. In fact, the acceptance of the ability of science to treat illness is so great that people look with disdain at anyone who uses any other techniques. However, people still rely on prayer when modern medicine still cannot provide a remedy.

Two Discoveries that Changed Modern Thinking

There are two discoveries, far more than any others, that have changed modern thinking. These two discoveries have been the center of the conflict between science and religion. The second discovery with regard to evolution is also having an impact on philosophical thought.

Galilio

The first discovery was the discovery by Galilio that the earth was not the center of the universe. Galilio proposed that the Sun, not the Earth is the center of our solar system. Astronomers believed incorrectly that the Sun was the center of the universe as well as the solar system for hundreds of years after Galileo's discovery. But, Galileo's original discovery was more significant, because it relied on scientific investigation, rather than religion, to provide an explanation for things in the physical world.

Although Galileo's conviction for teaching that the earth moves around the Sun was not technically a direct challenge to the infallibility of the Pope, the proceedings at the time, and for many years since, provide an example of the problem religion has with providing higher meaning. In order for people to rely on religion as a source of higher meaning religious teaching must be presented as being handed down from God or being absolute in some way. Such a position does not provide much flexibility over time. When people become aware of new information, that is in conflict with religious teachings and the new information appears to be on the side of science, religious teachings change slowly over time to accommodate this new information.

In Galileo's case he was convicted for publishing his *Dialogue*, and was placed under house arrest for his crime. His book was immediately placed on the index of Prohibited Books. Although it took almost 200 years, the Catholic Church eventually recognized Galileo's discovery as being correct, by taking his *Dialogue* off its index of Prohibited Books and allowed the teaching of the current understanding of the motion of the planets around the Sun.

According to the church, Galileo was tried and sentenced by the Holy office of the Inquisition, not by the church. And even thought Pope Paul V approved the Edict of 1616, which was against the Copernican theory that the Sum was the center of the solar system, and Pope Urban VIII condoned Galileo's conviction, neither pope invoked Papal infallibility. In fact the power of infallibility was not formally defined until Vatican Council I in 1869 - 70.

The real significance of Galileo's discovery and subsequent discoveries about the earth's position in the universe is that the earth is a very small place. This fact makes it harder to believe that people on earth are the sole purpose of creation. While Galileo's discovery may have challenged the authority of the Catholic Church, and his discovery also diminished to position of the earth in the universe, it did not have much impact on the belief that the universe was the work of some higher power. The discoveries of Charles Darwin and his theory of evolution do raise issues with regard to creation by a higher power. A close examination of Darwin's theory and the subsequent work of evolutionary scientists, also challenges much of philosophical thought, since the beginning of recorded history. While most people think of the theory of evolution as an explanation of how all life evolved on earth, it is also a theory about the process of development and adaptation which has significance, when thinking about higher meaning and purpose.

Darwin

Darwin's theory of evolution challenges religious teachings, because it challenges the teaching that humans were created by God and that we are separate and apart from the rest of creation. According to the Bible, God created the earth and gave man dominion over it. Since Darwin's theory of evolution directly effects us as individuals it is difficult to accept the full implications of evolution regardless of religious teachings. However, since science has advanced to a point where it can use the records stored in fossils from various time periods to prove the theory of evolution, it is difficult to argue with the facts that at least our bodies did evolve from other species. Most scientists believe that our large brain size, which provides us with the intelligence to be aware or conscious and communicate with others in our species, is also the result of evolution. While most scientists say that they do not believe in a God, in the sense of a religious God, some scientists believe in some natural God or some God of order. Nature appears to be orderly. Thus, it would appear that life was created according to some design even if it was not created by some higher, all powerful being.

A close examination of Darwinian evolution now can be used to support an argument that life on earth could have evolved without a plan, and that the human species is, as Stephen Jay Gould has characterized, just a "glorious evolutionary accident". The implications of believing that the human species is just an evolutionary accident is -- if it were possible to turn the clock back and run the evolutionary process forward again, there is very little possibility that the process would result in the human species again.

Since this concept touches the human psyche so deeply, it is a concept that is difficult even for scientists to accept. In fact, some scientist have suggested that this concept is so threatening to a person's sense of meaning that even scientists, who do not believe in God, still hold onto the belief, either consciously or unconsciously, that the human species did develop according to some plan.

In his book *Darwin's Dangerous Idea*, Daniel C. Dennett methodically and quite lucidly explains the implications of Darwin's theory of evolution from a philosopher's point of view. The complete implications of Darwin's idea are so significant that Dennett calls it "**Universal Acid**" and he writes over 500 pages to fully develop and document his ideas.

Dennett's analysis is so significant to higher meaning, in fact the subtitle of his book is *Evolution and the Meanings of Life*, that I believe it is essential to provide some overview of his ideas. I hope that I do them justice. However, I suggest that one must read Dennett's book, if you need to be convinced of the truth in his arguments.

Universal Acid

Dennett starts his analysis by describing the problems one would have, if one created an imaginary substance he calls universal acid. Unlike other acids, which only change some substances, universal acid changes everything that it comes in contact with. This compares with many acids that are highly corrosive to metal, but can be contained in glass containers, because they will not react with glass.

The quandary of developing universal acid is, once you develop it, how could you contain it. If you did develop universal acid, the ultimate impact would be that you would change all things on earth as it came in contact with the universal acid. The idea is somewhat like the China Syndrome, where the meltdown of a nuclear reactor would go through the earth until it reached China. By the same token, once Darwin developed his theory of evolution it was destined to change all philosophical thinking that came before.

It is also important to point out that based on other accounts of the how Darwin went about developing his theory of evolution, it appears that Darwin had no prior intentions of contradicting contemporary philosophy or religious teachings. His indent was only to publish what the facts revealed about the development of life. In reality, he struggled with the implications of his theory and held back on the publications of his work for many years, and when he did publish, he published only some of his ideas.

Prior to Darwin, world views were all of a top down orientation. I might add that for practical purposes, our daily lives today are still oriented in a top down orientation. All things fit into a hierarchy, which Dennett calls a Cosmic Pyramid. At the bottom of the pyramid is nothingness or empty space or not even space if you think like a physicist. Next comes chaos, which can be put in order. The order is created according to some design or plan. The plan is developed by a mind. And, the ultimate mind, the all powerful all knowing, mind is God. I might add that this is how I was taught in

Catholic school, that there is a God, and why he has the qualities that he has.

For his analysis, Dennett believes that it is important to distinguish between order and design, which at one time may not have been an important destination. For example, science has been able to determine the universe has order, but it has not been able to determine that it has a purpose. When humans make things, they make them for a purpose. When we make an automobile its purpose is to provide transportation. However, when we look at the universe through science, we cannot determine that the universe has a purpose.

What Darwin was able to show with his theory of evolution was -- if we have a universe that just has order, given enough time, design can flow from order and the rest of the Cosmic Pyramid is not required. Simply stated, Darwin was able to provide a new way of thinking that was originally criticized as a "strange inversion of reasoning". Darwin was able to overturn the Mind first way of thinking, which was essential to the proof that the universe was the product of Gods intelligence or Mind.

The anthropomorphic principle is the idea that we as humans think about God in human terms. Thus, we tend to give him human qualities only that they are the ultimate state of these qualities. Naturally, it is logical for us to assume that some ultimate power or creator would create the universe in the same way that a human would design and create the universe, if only we had the power. The power of Darwin's discovery is that life on earth has developed in a different way than the way a human would develop life.

It is this concept that has been the source of all the controversy over evolution and not the simple fact that men are descendants from the apes. Ironically, because Darwin's ideas were so controversial many people tried to disprove his thinking or find some fatal flaw. However, these efforts to disprove Darwin as well as much of what evolutionary science has done since have only reinforced his work. Darwin also left open the possibility of reconciling his ideas with prior world views. He suggested that while the process of natural selection was an automatic process that flowed from order, the design of the automatic process came from the Mind of God. However, this was more an accommodation to his critics rather than a true limitation to his ideas.

Design Without Intelligence

Until Darwin's discovery, the only design that we had been able to discover was design through intelligence. We are all familiar with the process of research and development, which is really a two step process. We define the term pure research as the process of discovering basic laws of how things in nature interact. Using this basic information as a resource we then design an artifact to perform a specific function. The essential part of the process is that we use our intelligence plus information about nature to configure the artifact so that it uses the basic forces of nature to accomplish a purpose that we have established at the beginning of the R&D process.

If one looks at the complexity and elegance of nature, and one also only has one concept of design, it is not hard to conclude that something as complex as a human being must have been made by some higher intelligence. Most people have heard the criticism of evolution, which addresses the process of random variation and natural selection, which states -- "No matter how long a thousand monkeys banged on typewriters they would never be able to create even one page of meaningful text".

What people do not understand is that Darwin discovered a whole new way of design. He combined an algorithm working over a long period of time with a means of storing the results of each step of the algorithm so that it did not have to be done over again each time. This process is what Dennett calls "the accumulation of design".

Although I am trying to make this explanation as bare bones as possible, I do have to go back and define some terms. An algorithm is a process, which does not require the input of intelligence while the algorithm is working. It also could be described as what we think about as a mechanical process. The algorithm that works in evolution is selective adaptation of a species to the environment. The device that stores the information from each step is DNA. It is interesting to note that at the time that Darwin made his discovery, he was not aware of genetics, although Mendel had already done work in this area.

It is also important to think about the way we think about an artifact and the design process. When individual humans take an active part in the design process, they have the end purpose of the artifact at the beginning of the design process. I would like to emphasize the word individuals here, because as I will discuss later, I believe that if you look at the long term development of a product in the market place, a product actually follows a Darwinian design process.

In business, it is common for a company to reverse engineer the products of their competitors. In reverse engineering, a company already knows the function of the product. What it is trying to determine is how the product was made, or how much it might cost to make the product. One of the mistakes we make when we look at evolution through the fossil records is we try to apply reverse engineering to the process of evolution.

Although there have been millions of steps in the evolutionary process that resulted in the development of the human species to date, the process is only on the current step. The process has no way of knowing what the next step will be until variation occurs and the one species that is more adept at survival in the environment, actually prevails over other species. Approximately 30,000 years ago Neanderthal men could no longer compete with Homo sapiens or Cro-Magnon men and they became extinct.

When people look at the human species or any other animal and its ability to function in its environment, it is easy to assume the evolution is a very efficient process, when in fact it is just the opposite, however, it is very elegant. Many species have vestiges in their anatomical design of capabilities that are no longer required in the current environment. For example, some whales are born today with small leg type limbs, which are holdovers from the time, about fifty million years ago, when the whale's ancestors walked on land.

The evolutionary design process has limitation in the sense that once a species moves down a particular path, through evolution, it may never be able to develop a curtain capability. For example, it would be almost impossible for plants to develop eyes or legs. We also are only able to observe the species that survived long enough to leave a record in the fossil record. Evolution involves a constant process of emergence and extinction where species are relatively short lived. Thus, one could say the process is very inefficient, at least compared to the human method of design. However, if survival of life under conditions that are not known in advance is the end objective, this method of design appears to be the most reliable, if not the only process to use.

While most people do not have a problem accepting Darwinian evolution as an explanation for some of the development of life, many people still have a problem accepting the idea that evolution is responsible for the entire process from emergence of the first complex molecules all the way to the human species.

Dennett introduces the idea of cranes and skyhooks. In his analysis, a crane is some acceptable step or improvement that can be explained by observed data or logical thinking, whereas a skyhook is some conceptual force or factor that is beyond logical thinking or the observable data. In evolution, a skyhook could be the hand of God. Since Dennett devotes a major portion to an analysis that supports the idea that Darwinian evolution can be explained through the use of cranes alone, I will simply state here that he makes a convincing case.

However, in terms of the meaning of life, I believe it is important to look at the idea of cranes. More important, this example may illustrate why there is higher meaning to life, which can be use to establish personal meaning or a least can be used as guides in living life.

We Build Upon What Came Before

Darwin's process of evolution takes place by the smallest possible variation in design that is then selected by environmental factors. This is a slow process, so slow in fact, that it can not be perceived in action with the exception of observing some organisms, such as viruses, that have a very short time frame for each generation. There are some cranes such as sexual reproduction and human intelligence that can accelerate the process. These cranes do come at a cost. In the case of sexual reproduction only half of an organism's genes are passed along in each generation and the organism must find a suitable partner. With human intelligence, which can anticipate rather than just react to current circumstance, we are able to develop genetic engineering. Through genetic engineering we can develop species that wouldn't develop under natural circumstances.

What we are able to accomplish as individuals or even as a species depends upon a very long period of design work. It has been about 600 million years since the first multicell organisms emerged. While we may be able to make huge leaps in design in terms of historical evolution we are making only a small addition in terms of the design work that has come before. While we believe that we have made huge advances in technology in the last hundred years, we have only learned how to use tools that were developed over billions of years. We did not create the tools we are simply using them.

The timing of the development of the theory of evolution may be of interest to people as they think about the meaning of life. When Darwin published ***On the Origin of the Species by Natural Selection***, the English class system dominated society. People in the upper class believed that their superior position in society was due to some inherent qualities that they had at birth. It was hard for the upper class to accept that all people, including primitive peoples, had the same physical and mental capabilities. To the English gentlemen people sitting around at afternoon tea required more intelligence than people sitting around a campfire in the African Bush.

Higher Meaning Changes

It would be easy to assume that if there is a higher meaning to life that it would be absolute and unchanging. Religion does not hold the position that it did several hundred years ago, because science has shown some teachings of religion were not in sync with discovery. Also, as a means of social control, religion has to change with changes in society. History has shown that it has in fact changed dramatically over time.

People may think that our lives are not impacted by philosophy, but at many levels many thought processes, which we refer to as intuitive thinking, are based in philosophy. I believe that Daniel Dennett has shown that Darwin's discoveries makes us reevaluate what we think about as true by intuition. Darwin is not the only scientist to make us aware that reality is not what it may seem. Einstein's theories about space and time also propose ideas about reality that appear illogical compared to everyday experience. However, Einstein's theories about space and time are about situations that will never be part of our everyday experience. Darwin's discoveries relate to our very existence as a species.

Many people challenge Darwinian evolution because it challenges other sources of higher meaning in their lives. The challenge wouldn't be so important if it did not relate back to how individuals see their place in the universe and the meaning of their lives. However, I would disagree with Existentialists that life has no higher meaning simply because life may not have been created by a higher power according to some grand design. Personal meaning will always have some reference to some understanding of a reality broader than the individual. The understanding that the human species is the result of design created through the evolutionary process is far from a world in chaos.

People seek higher meaning in a changing world because it is human nature to seek stability. However, people can stand comfortably on a moving train. Higher meaning can change and still provide stability, and with the rapid advance in human knowledge people should expect some change in higher meaning as they go through life. People also want their lives to mean more than their daily existence. They get comfort in the idea that they are part of God's grand design. However this must be accepted on faith, since by definition it is beyond proof. Personally, I get more comfort in understanding my place in the universe without having to rely on faith.

First Higher Meaning

It is so obvious that it seems like it is not worth mentioning that our first concept of higher meaning comes from our parents. Besides providing the details of higher meaning, such as the details of a religious faith, we are taught that life has meaning and that everything makes sense. For many people the higher meaning that we get from our parents is the only higher meaning that we have for the rest of our life. Although people, as they grow to become adults will challenge some of the details of higher meaning in the process of establishing an identity, only some people will challenge everything. For example, people may move from participating in an organized religion to believing in a personal God. Such a person still believes in some higher power in the universe, but the person does not necessarily accept all the moral values that are taught by the formal religion into which they were born.

Many scientists today are either atheists when asked about a religious God, or agnostics even when it comes to a personal God. I believe the reluctance to come to a decision one way or another is the result of the connection of higher meaning to morality. Most likely, religion was a more important part of their lives at one time. At that time, religion provided both higher meaning and a guide to morality. It is hard to separate higher meaning from morality. As long as we are part of society we need the morality to govern our behavior. Therefore, we hang on to the higher meaning if only to support the morality.

Like many other topics touched in this book, one could devote at least a full chapter if not a whole book on the sources of morality. However, I would like to maintain the focus on higher meaning. But, it is important for a discussion of higher meaning to at least point out that higher meaning and morality are tied closely together, and that ethics has always been one of the prime subjects of philosophers.

One of the factors that defines a social group are the values shared by that group. One of the roles of a parent, if not an absolute duty, is to pass on values to their children. In the case of some values these values are so integrated with a person's identity that a rejection of that person's values is the same as the rejection of the person.

In the chapter on work I will talk about the value of hard work, which we commonly refer to the work ethic. Although the work ethic is a concept that was never really proved to be true, the work ethic is a value that has been passed from generation to generation. There are many fathers who would view their son's rejection of the work ethic as a personal rejection. By the same token, acceptance of a specific higher meaning for life is so engrained in society that one cannot reject that higher meaning and still participate in the society.

In the decade of the 60's, a significant segment of the young people rejected their parents values, which included an emphasis on consumerism and the excesses of the commercial sector in America. These excesses were also viewed as taking too large a toll on our environment. To some of these people it was viewed as necessary to live apart from such a society in communes either in the city or out in the country. While other people may question the higher meaning they get from society, they are not willing to pay the price of social separation in order to follow a belief in some other higher meaning.

Death

The concept of higher meaning is important when people talk about or consider death. The human mind seems to naturally operate with the concept of forever. Although we all know that we are going to die someday we go through our lives almost to the very end without any thought of death. Naturally the end of our existence is not a very pleasant thought.

Throughout most of recorded history, life for the average person has involved a lot of suffering and discomfort. Not until the twentieth century did anyone live completely free from exposure to the elements. People lived in houses without central heating, and many people worked outdoors. Most people worked at manual labor and children started to do some work either on the farm or in the trades by the age of ten. Thus, it was appealing to look forward to a life after death where everyone was free from this suffering.

We still live in a system that operates on the principles of reward and punishment. We get rewarded for positive efforts, which today includes mostly work, and we get punished for unsociable behavior. This system is much more effective if it can be tied to some system of ultimate reward and punishment. This system of reward and punishment for good and evil is one of the greatest sources of loss of religious faith. If people are rewarded for good behavior we always have the question, why do bad things happen to good people. By the same token, religion has to explain why people do not get what they pray for. Religion's stock answer is that we don't know what the true good is we are seeking.

If we can think in terms of some higher meaning in life that involves some existence after this life it is comforting for us both when we think about our own death and the death of others. The meaning of our lives are thought about in terms of accomplishments and relationships. If there is nothing after this life all our striving and work toward goals comes to an end. Also all our relationships end.

Unfortunately, when most people die their life is not completed. It is better to think about it as interrupted. On the day we die we still have relationships with friends and relatives and we usually have some activities that we could carry on if we were given a few more weeks, months, or even years to live.

A higher meaning for life also provides comfort for people when they loose loved ones. It is comforting to believe that people will rejoin people who have died when they themselves die. Especially when people die at an early age, these people are viewed as going to an early reward.

Fortunately, for society the will to live is a very strong instinct. Otherwise, society would have a problem with maintaining the species. Prohibitions against murder and suicide are essential to maintaining the species as well as protecting the rights of the individual. These prohibitions are one area of religion where higher meaning must be directly linked to social control.

The one time when the impact of death can be somewhat mitigated is when someone dies for a cause. Although all wars are senseless, and religious wars seem to be the ultimate contradiction, people who die for God or country seem to cheat death of its ability to take away meaning. It is important to keep this concept in mind when thinking about terrorists that are willing to die for their cause. Modern day terrorists are not the only people who consciously and actively seek to die for a cause. Early Christianity had to contend with people who were more than willing to become martyrs. The church eventually had to prohibit people from seeking to be martyrs.

The Quandary of Meaning

While no one seems to be able to determine with certainty what the ultimate meaning of life is, most people have a strong need to have a meaning for life. Maybe someday science will discover that when humans engage in activities, which they perceive to be meaningful, their brains produce some chemical, which either directly produces a sense of well being or is a response promoted by some genetic programming that causes us to engage in meaningful and possibly ethical behavior.

We already accept such explanations for our desire to eat and engage in sexual activity, and we freely acknowledge that we share these chemically motivated biological activities with other species. On the other hand, the human species seems to be the only species that has a need to engage in purposeful activities. Eating and sexual intercourse are meaningful activities in the sense that they are for the purpose of maintaining the individual and the species, but humans like other species can engage in these activities without any thought of the end purpose or goal that these activities will achieve.

In fact, in the case of sexual intercourse the biological purpose of the activity could be different from the human meaning of the activity. The biological purpose of sexual intercourse is reproduction of the species, while the meaning of a specific act of sexual intercourse can range from an act of affection or play to an act of violence, in the case of rape.

In the time frame of evolution, human meaning or conscious purpose is a very recent thing. As can be noted from the family tree of primates in Figure 3.1 below the human species emerged sometime between five and ten million years ago, which is after the gorilla emerged as a species and before the chimps emerged. If you think of our bodies as engineering projects, which they very much are, we have the benefit of all the engineering in a gorilla's genes before they started adding unique features to their DNA. On the other hand, chimps have the benefit of all the engineering that we acquired through evolution before they became their own species.

Scientists believe that our ability for conscious thought and our ability to communicate through speech is a function of our very large brain size relative to our body mass. Whales have brains that are four times as large as a human brain, but their body mass can be as much as 1000 times that of a human. Other species have the ability to communicate but our capability is far more sophisticated. Our ability for speech emerged with Neanderthal man about 120,000 years ago, and our large brain size emerged with Homo sapiens, about forty thousand years ago.

As recently as a few hundred years ago we knew very little about the origin of the human species or the age of the universe. For example, it was commonly believed that the universe was only about 10,000 years old. It would be logical to assume that the earth was only 10,000 years old, if any recorded information, included verbally transmitted information, was only about 10,000 years old. There is little question that throughout the period of recorded history the human species was the dominant species on earth. Although we knew that we were the dominant species we had no knowledge of why we could possibly be in the position we occupy in nature.

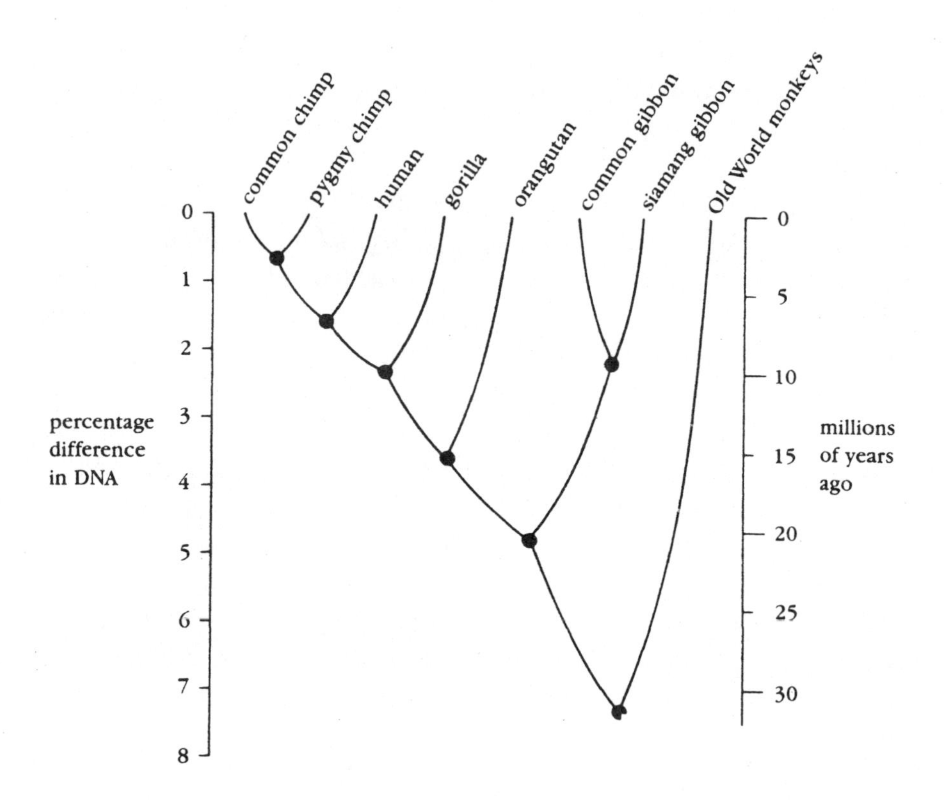

Family tree of higher primates. Trace back each pair of modern higher primates to the dot connecting them. The numbers to the left then give the percentage difference between the DNAs of those modern primates, while the numbers to the right give the estimated number of millions of years ago since they last shared a common ancestor. For example, the common and pygmy chimps differ in about .7% of the NNA and diverged around three million years ago; we differ in 1.6% of our DNA from either chimp and diverged from their common ancestor seven millio ago; and gorillas differ in about 2.3% of their DNA from us or chimps and diverged from the common ancestor leading to us and the chimps around 10 million years ago. [Diamond 1992]

Figure 3.1

Artifacts and Memes

The human species is the first species that we are aware of that is able to create artifacts and memes. According to archeological findings, Neanderthal man created crude tools. Artifacts, which could also be thought of as tools, are anything that are created with the view for subsequent use. They are something we make for a purpose. Thus, it could be said that Neanderthal man had sufficient intelligence, in spite of his smaller brain size to engage in purposeful activity. We can only speculate, but I believe that Homo sapiens were the first species to create memes.

Richard Dawkins, an evolutionary biologists, coined the term meme to describe the most basic piece of a concept that still retains the essence of a concept. He also called it "a unit of cultural transmission or a unit of imitation."

> "Examples of memes are tunes, ideas, catch phrases, clothes fashions, ways of making pots or of building arches. Just as genes propagate themselves in a gene pool by leaping from body to body via sperm or eggs, so memes propagate themselves in the meme pool by leaping from brain to brain via a process which, in a broad sense, can be called imitation. If a scientists hears, or reads about a good idea, he passes it on to his colleagues and students. He mentions it in articles and his lectures. If the idea catches on, it can be said to propagate itself spreading from brain to brain." (Dawkins 1976, p206)

Thus, Homo-sapiens are the first humans to have the capability to think about the meaning of life. Another way of looking at this would be to say the concept of meaning only has functionality when we are dealing with Homo-sapiens or present day man. It could be said that the problem of meaning is a relative new problem in terms of the history of life on earth.

I would disagree with the existentialists that science has shown us that life has no higher meaning, or better yet that life does not have the meaning we get from religion, which we believe only on faith.

Science has been able to discover what is. More importantly, science has been able to discover why we seek meaning. Science may not be able to endorse or support the meaning that we once got from religion, but that does not mean that life has no meaning.

Even existentialism believes that personal meaning is less of a problem philosophically. Seeking higher meaning is only a problem for some people. According to surveys, most people are satisfied with the higher meaning that they are able to get from religion, personal introspection, or some other source. The time frames of science that relate to a higher meaning of life are at best impractical for the individual. A single generation is no more than a heartbeat in terms of the history of the human species. Thus, in chapter six we will look at how an individual finds meaning in life and what factors make that experience worth having. On an individual basis, the meaning of life simply has to provide purpose to one person -- the person asking is life worth it or am I worth it.

Chapter Four

Life Is Unpredictable and Everchanging

Life is what happens to you while you are making plans to do something else.

Physics Large and Small

One of the first branches of science to study nature is physics. Naturally, the first things to be studied were the things that we experience in everyday life. Some of the first laws of physics, which are also most familiar to the average person, are Newtonian mechanics. Later, when tools were developed to study nature on the scale of atoms and molecules, new laws were developed such as the laws of quantum mechanics. The combination of the theory of relativity and quantum mechanics is often referred to as modern physics.

The most important thing that science has learned from quantum mechanics is that nature behaves differently at the level of atoms than at the level of baseballs and planets. When scientists started to develop the laws of quantum mechanics, the nature they observed was often contrary to what we experience at the human-size level. Some people, such as Albert Einstein, could not accept nature as observed and attributed the contrary observation to some additional factors that had yet to be discovered.

One of the most well-known laws of Newtonian mechanics says "for every action there is an equal and opposite reaction." We experience this law when we walk along the street or drive a car. While walking, we lean forward and push against the ground to move forward. In a car, a small amount of gasoline in the engine explodes and the energy created pushes a piston down. Through gears and shafts, the energy is transferred to the wheels, which also push against the road to move the car forward.

The laws of Newtonian mechanics also govern such familiar things as the tides and the motion of the moon around the earth and the earth around the sun. The common thread in all laws of Newtonian mechanics is that we can predict what will happen, if we are given the facts with regard to a situation. These laws seem to follow what some people might call common sense—they are a formal way of explaining what we experience.

When we think about the elementary particles—electrons, neutrons, and protons that make up atoms—we have to change our mechanical intuition of our day-to-day experience. An electron moving around an atom is not like a baseball when it is hit by a bat. The baseball has a specific trajectory, while the atom has many possible, simultaneous trajectories. When physicists started to develop the laws of quantum mechanics, they were required to deal for the first time with the probability that an event will occur and it will be affected by the randomness of nature. In other words, we know that things will happen, but we cannot always predict when they will happen—even when we have all the facts. As Albert Einstein said, "I can't imagine God playing dice with the universe." This quotation illustrates Einstein's objections to the theory of quantum mechanics.

Randomness in Nature

The decomposition of radioactive material is another example of the randomness of nature. Scientists can determine that a particular element will decay or change from one form to another at a certain rate. The decay is known as a half-life or the time that it takes half of the radioactive material to change. While scientists are certain that one half of the atoms in radioactive material will decay in a certain period of time, usually thousands of years, they cannot predict when any one particular atom will decay. The atom could decay today, or it could decay ten million years from now.

There are many other areas in science in which randomness is an important part of the study of the way things are. Take the study of fractals, for example. The law of fractals is at work in snowflakes—while every snowflake has six points, no two snowflakes are exactly alike. In fact, random variation can be observed throughout nature. Besides snowflakes, fractals determine how leaves grow on trees and ripples form on a stream. We accept the variations in people, because human beings are complex organisms. However, we don't expect to see this in snowflakes. They are simply a collection of water crystals. The formation of water crystals is determined by a relatively small number of physical laws.

Individuals and society as a whole also live under this dichotomy of prediction and uncertainty. The average life expectancy in the United States for males in 2001 was 77.6 years. However, as we all know, some males will live only a few days, while other males could live to be more than one hundred years old. By the same token, studies have shown that smoking shortens the average life expectancy and that more education leads to higher average annual incomes. But some smokers die of cancer, while others don't. And some people with little formal education earn far more than people with advanced degrees.

You could point out that in most cases more than one factor is involved in life expectancy and a person's annual earnings, and you would be right. We do live in a chaotic world, and there are many factors that affect our lives. However, some of them are under our control, such as choosing not to smoke or spending more time in school to get more education, while others are not. There are many other random events that we have no control over but do have a significant impact on our lives.

Scientists have recently determined the human race owes its very existence to one random event that occurred about sixty-five million years ago. Up until then, dinosaurs were the dominant species on the planet earth. These creatures, which were well adapted to survival in their environment, dominated the earth for about one hundred million years. Mammals also existed at that same time, but due to the dominance of dinosaurs, they were not able to advance beyond small rodents.

Then a random event occurred—the impact of an asteroid in the Yucatan peninsula. The environment changed rapidly. As a result, there was a mass extinction of about 85 percent of the species alive at the time. Fortunately for humans, small rodents (our predecessors) were able to survive and, without competition from larger animals, were able over time to slowly evolve into larger animals. That was how the human species was able to emerge from the evolutionary tree.

In fact, one of the factors in evolution is random change in a species that makes it more adaptable—better able to live in the environment as it changes and to compete with other species that have not adapted as well to the current environment. You might agree that random events occur, but maintain that they do not happen frequently enough to have an impact on your life. Think again.

The Big Things in Life Are the Result of Chance

Take a minute to examine the significant events in your life: birth, marriage (if you are married), career (how you chose a career and where you went to work), serious injuries or illnesses, the death of family or loved ones. Consider the circumstances around these events.

Our lives are also determined by some random laws of nature, which I call the quantum laws of life. These laws are stated as follows:

- Most of the important things in our lives are determined mainly by chance—we don't have any control over them.
- While we may think we have control over our lives, we are mostly going with the flow; that's what makes life so much fun.
- Some things that work for society as a whole may not work for the individual.
- Some small things not under our control can change the course of events for or against us.
- Justice may be blind, but so is fate. Good things happen to bad people, and bad things happen to good people.
- We can control our actions, but we cannot control the universe. The universe does not reward or punish us based on our actions. However, society may.

From the point of conception going forward, our life is governed by chance. During the act of sexual intercourse, millions of sperm cells are released. About half of these cells contain female chromosomes, and the other half has male chromosomes. If any of these cells are successful in combining with an egg in a woman's uterus, only one will determine the sex of the baby that is eventually born. Chance determines if a person is born male or female and able to take advantage of the rights and opportunities afforded to each gender. Living your life as a female can be significantly different than living it as male. For one thing, only women can experience pregnancy and childbirth. On the other hand, women in some societies do not enjoy the same rights and opportunities that men do.

We do not choose our parents. We have no choice whether we are born into a wealthy or poor family, into a well-adjusted or dysfunctional one. However, our parents provide us with their financial and emotional support for at least twenty years of our life. They provide us with our initial set of values and goals in life, either through words or actions. If our parents provide a good role model, there is a good chance that we will follow in their footsteps, regardless of what we might inherit from them in monetary terms. For those who are fortunate to be born in a well-connected family, the social connections alone can be enough to almost guarantee a comfortable life regardless of effort or basic abilities.

At least in the United States, the person we marry is mostly dependent on chance. A man or a woman will most likely marry someone with similar social and educational background, but the specific person that you meet is open to almost infinite possibilities. The chance meeting under unusual circumstances is the theme of countless romance novels and movies.

Work fills a major part of our day, and we do get to choose what we do for a living. Even people with little education and limited opportunities have some ability to choose what they do to make whatever money they can. The alternatives may not be attractive, but we do have a choice. However, even people with the best jobs can be subject to the laws of chance.

Choosing an employer is somewhat of a random process. We only can apply for jobs that are available at the time we are seeking new or improved employment. The more important the job, the fewer opportunities are available. There is only one position as president of the United States; there are only one hundred seats in the U.S. Senate; there are limited job openings as CEO of a Fortune 500 company.

Working for a company that launches a trend setting product or develops a major lifesaving drug can be mostly a factor of chance. As a job applicant, we are not privy to what product breakthroughs a company has in the pipeline. We have to choose an employer based on its previous track record and its drive to be creative and hope we are lucky enough to be a part of some groundbreaking projects.

Chance Affects Everyone

Chance can affect everyone—from a young man just starting out in life to the president of the United States. On the weekend of July 4, 2001, two young men were playing with a Frisbee on the beach in Island Beach State Park in New Jersey. While it was only a partly cloudy day, they were both hit by lightning. According to the weather service that day, there was a developing thunderstorm two miles off shore that was responsible for what might be characterized as a freak lightning strike. One man died and the other lived. That they should be struck at all was completely random, and then that one man should live and the other die was equally as random.

The presidential race between George W. Bush and Al Gore was decided by what could also be described as a chance event. After all the money and effort spent on the campaign, the election was eventually decided by a butterfly ballot. One of the items disputed in the weeks following this close election was the design of a butterfly ballot used in certain precincts in Florida. The way in which the ballot was designed made the ballot potentially confusing to voters. Several voters who wanted to vote for Gore complained that they had inadvertently voted for the wrong candidate because they were confused by the design of the ballot. The number of votes in question was considerable—enough to give Gore the margin of victory. While Democrats cried foul, Republicans pointed out that the ballot was designed by an election official who was a Democrat.

Not only were the lives of George W. Bush and Al Gore significantly changed by this random event, but also the lives of all the people in the United States will feel the impact of this random event to some extent for at least the next four years.

We look back at events, and we see how random events can have a major impact on our lives. This makes us even more determined to figure out this thing called randomness. Through science, we have been able to learn why major geological events, such as earthquakes and volcanic eruptions, occur. However, we are still a long way from being able to reliably predict these events. From past experience, we have learned that large volcanic eruptions occur as frequently as every few hundred years and that usually the immediate impact of these events is less significant than their secondary impact. For example, while the eruption could cause loss of life in the thousands, the impact it has on weather can lead to future crop failures and the deaths of millions.

Random Events Make Life Interesting

While we may think we have control over our life, we are mostly going with the flow. To a degree, the randomness of life is what makes life interesting. Change can be stressful, so the amount of change must be within limits. Everyone has a different tolerance for change, but a life without any change is boring. We go on vacation for a change of scene and to experience new things. We read novels and go to the movies to experience things that we do not have in our ordinary lives.

The popularity of lotteries shows that people believe that what does not work for people as a group does work for the individual. The odds of winning a lottery are small. Lotteries are, in fact, a losing proposition for everyone as a whole, because the lottery does not pay out as much as it takes in.

We also seem to think that lotteries provide a way of being fair. We employ lotteries in situations in which only a certain number of people can be chosen and money or some other factor cannot be used to make a selection. A lottery was used for the Selective Service draft, although it does seem a contradiction in terminology.

A coin toss is used at the beginning of a football game to determine which team gets to receive the ball. Lotteries are used to give landing assignments at crowded airports.

Much of the things that give us satisfaction in life and many of the things that give us grief are not under our control. Unfortunately, the value system under which we are raised as children and which is later reinforced by society tells us that we are responsible for the things that happen to us. Good things happen to us if we are good, and bad things happen to us when we are bad. It is the American way that hard work pays off. The things that we are able to acquire in life are a reflection of how good we are or how hard we have worked. While I am not trying to discourage initiative, nor am I encouraging immorality, I believe that a different perspective on our accomplishments and failures will make our life more meaningful as individuals and will help us make important choices as we move forward as a society in the twenty-first century.

Hope

People often ask why bad things happen to good people. After all, religious teachings tell us that we will be rewarded for good deeds and punished for our sins. Thus, when good people experience bad things, it seems to go against the justice that we expect from God. We are given various explanations, such as God is testing our faith, to reconcile this inconsistency and to maintain hope.

But I suggest that the justice of God theory is really an instrument of social control to influence our behavior as social beings. To accept the fact that the universe is random does not have to lead to despair. Every time someone prays they are not assured that there prayers will be answered, but they have hope. Sometimes it appears that our prayers are answered. However, the answer to prayers could also be explained by the randomness of nature.

Medical science has shown repeatedly that a patient's mental attitude toward his or her chances of recovery can have an impact on recovery, all other things being equal. While the source of hope is different when we provide a natural explanation for a favorable outcome, hope is still present.

Life Is a Journey, Not a Destination

There are countless examples of people who have worked hard to achieve certain goals and then experienced a significant letdown, if not outright depression, after they had achieved their goals. It has been said that some of the first astronauts become depressed after they achieved their goal of going to the moon.

The idea that life is working toward a destination probably comes from the religious idea that our goal in life is to work toward some form of perfection either in this life or the next. The added advantage of working toward another life after this one could be an escape from the stresses and pain of this life.

We learn from evolution that life is constantly changing, but not necessarily leading to any goal other than being best adapted to the conditions that exist in the environment at the moment. The lesson that evolution tells us is that the essence in life is the journey. The journey, if experienced properly, provides the joy of experiencing the endlessly unfolding surprise.

The concept that life is a journey, not a destination, is expressed in many different ways. A common refrain, on sports fields and off, is that it is not important whether you win or lose but how you play the game. The idea of good sportsmanship is implied in this statement, but who would want to play any game, if they knew what the final score was going to be in advance? Even when two teams may be mismatched, there is always the hope for an upset, for the underdog to emerge victorious (or least not get beaten as badly as expected going into the game).

While many people desire to have great wealth, people who inherit wealth often do not choose to live a life of leisure. I believe that these people choose to pursue careers and work not out of a sense of moral obligation, but to expose themselves to additional possibilities in life.

For people who are born into ordinary means, randomness or the lack of certain outcome provides a great incentive to make the best of life. In societies such as that in the United States, the pursuit of happiness is dependent upon freedom. Even the average man will be happier if he feels that he did not have to work under any restrictions.

Order in Chaos

With so many things subject to random events, it is easy to fall into despair that the universe is nothing but chaos. While we have shown that indeed many things appear to be chaotic, we may learn some day that there is, in fact, some deeper order that can account for what appears to be chaos.

Much recent scientific study has attempted to explain the apparent chaos in the universe, although to some it may sound more like science fiction than science. For example, some scientists suggest that we live in only one of many parallel universes. They theorize that we are experiencing only one of the possible outcomes of a potential event in this universe and that all the other potential outcomes are actually being played out in other universes. These theories involve such exotic concepts as wormholes, which science speculates are connections between the parallel universes.

In recent years, some of the most exciting study in astrophysics has involved black holes. Black holes are assumed to be formed after certain stars collapse and form a mass so dense and with such a strong gravitational force that not even light can escape from its gravity. Some scientists speculate that black holes may be connections to other universes.

A new branch of science called chaos theory takes another approach. It is attempting to develop laws that try to put some order in things that do not appear to be organized. The basic principle of chaos theory is that while we may be able to explain the basic forces that affect things in isolation, once we combine several simple things together, it is impossible to predict the exact outcome of the things in combinations. One example often used to explain this phenomenon is the ripples on a stream as water flows over rocks. While there is some pattern to the water flowing over the rocks in that there are constant ripples, the pattern of these ripples is constantly changing as a result of various forces occurring at random intervals.

Patterns in weather are another example. The basic pattern of high and low pressure systems is easily explainable, but the specific occurrence of highs and lows and their intensity is more difficult to predict and explain in the short term and almost impossible to predict in long term of more than a few weeks.

These theories eventually lead us back to the basic philosophical problem that it is hard for us to study and understand things until we have the right tools. We must have tools that aid our senses sufficiently to first allow us to come to a complete understanding of these events, and later to use these tools in an experiment that allows us to prove a theory through consistent outcomes and empirical observation.

The QWERTY Keyboard

Sometimes one event that could be considered somewhat random starts a series of events that eventually produce an end result that is completely contrary to what one would expect looking at the situation from a purely objective point of view. Such events don't happen all the time, but they do occur frequently enough to be worthy of notice and to have an impact on our lives. The QWERTY typewriter keyboard is often used to illustrate this phenomenon.

The first six letters on the top row of a typewriter keyboard are QWERTY. Typists using the early mechanical typewriters eventually became so adept at their jobs that they could type faster than the machine could respond. This caused the keys to jam. For people who may have never seen a typewriter, all the keys on a mechanical typewriter lever up to the same point on the roller or platen where the character's image is stamped on the paper. Only one key can strike the exact same spot as the carriage moves the paper across that spot. If a bar for one character has not fallen away before the next bar from another character tries to occupy the same space, the bars get jammed together.

To slow the typist down, the keys were arranged in a less efficient pattern. When the Remington Sewing Machine Company started mass-producing typewriters, it adopted the less easily jammed QWERTY keyboard. Many typists learned to use the keyboard, and other manufacturers decided to use the same key layout, which in turn meant that many more people learned the layout. When electric typewriters and eventually computers were introduced, jamming keys was no longer a problem. However, despite significant efforts to adopt a more efficient keyboard, we still use the QWERTY keyboard today.

This phenomenon was labeled "increasing returns" by Brian Arthur. In other words, the product or practice that is already in place has a significant advantage over another—even though the new product or practice may be technically superior. Trends, such as the QWERTY keyboard, seem to have a momentum of their own. Such trends do not have an external explanation or valid reason for continuing—yet they do.

We also see this in the concept of those who become rich, or the rich get richer. Some people who appear to be self-made may have been just lucky at one time. These people, who were lucky to benefit from the confluence of events at one time, generated the economic means, early on, to further exploit the trend as it developed momentum. Looking at it another way, once other people recognize a trend they bid up the resources needed to exploit the trend. Later on, it costs more to get in. Thus, the returns are not as great proportionally.

Complexity and the Edge of Chaos

In spite of the chaos that we observe in the universe, we do experience order. In fact, we experience order in some very complex systems, but we don't yet have an explanation for this order. Some people attribute this order to a manifestation of the hand of God in the universe, but scientists are now trying to determine if there are some laws that will explain complex systems. If we can understand how complex systems work, we may be able to exercise more control over the economy or explain why certain technology develops, while other equally possible technology remains undeveloped. We may be able to use this science to explain how life itself developed more than four billion years ago.

Some people, who object to the theory of evolution, correctly point out that it is not possible to explain the development of the human species strictly through random evolutionary accidents. The study of complexity may reveal some self-organizing capability that will allow for the progress that we see recorded in the fossil record. Science cannot yet explain how our brains, which we know to be the most complex organisms on the planet, can give rise to such functions or qualities as feeling, purpose, and awareness. We know that brains contain billions of neurons and that each of these neurons can make connections with thousands of other neurons. Although the number of potential connections is finite, the number is so large that it is one thing in nature so close to infinity that it is not possible to conceive of the difference between that number and infinity.

Some scientists believe that the principles of quantum mechanics may some day explain how our brains function.

We take for granted the fact that the U.S. economy, in a period of less than four hundred years, has grown to dominate the world economy. We know that people try to satisfy their material needs by organizing themselves into an economy of buying and selling among people of specialized skills and resources, but no one seems to be organizing the American marketplace. Economists admit to fine-tuning the economy, but no one says what we should make or how we should make it. In fact, the very principles of the U.S. system provide for independent action.

In addition to being self-organizing, systems appear to adapt to the environment around them. Birds adapt to other birds around them to form a flock. Organisms constantly adapt to each other to form an ecosystem that allows the system to be viable in the current physical environment. They constantly try to turn what has happened to their advantage in the struggle to survive. Finally, all complex, self-organizing, adaptive systems have a dynamic quality that separates them from systems that are just as complex such as computer chips or snowflakes. They somehow acquire the ability to bring order and chaos into balance. This balancing point is referred to as the "edge of chaos." The edge of chaos is a region where parts of the system don't lock into place, but they also don't dissolve into complete chaos.

The edge of chaos is where innovation is constantly taking on the status quo and where even the most entrenched will be replaced if it no longer is appropriate to the current environment.

Characteristics of Complex Systems

There are several things that can be said about complex adaptive systems, such as brains, environments, economies, scientific communities, and political parties. John H. Holland, of the

University of Michigan, listed some of the key characteristics of complex adaptive systems in a presentation to the Santa Fe Institute.

1. Complex systems have many agents or parts working in parallel. In an economy, the agents would be firms. The control of these systems tends to be disbursed. In an economy, each firm does what is best for its stakeholders. The government may try to fine-tune things with tax policy or interest rates, but most of the productive activity is directed by firms.
2. Complex systems have many levels of organization. In an economy, the firms are grouped into industries. The firms themselves are organized into subsidiaries, divisions, departments, and work groups. In nature, the different levels are manifested in the various levels of predator and prey, while at the same time, there are many other symbiotic relationships. For example, certain birds eat the bugs that infest the hides of larger animals.

   As anyone who has worked for a major corporation knows, these agents are constantly rearranging their building blocks after they gain experience with their environment. In living things, this process is manifested by organisms as they modify and rearrange their tissue through the process of evolution.
3. Complex adaptive systems anticipate the future. In an economy, it is obvious that people anticipate the future. The working of the stock market is one of the best examples of people anticipating the future. However, anticipation is somewhat less obvious in nature. Learning from experience and genetic material, which predisposes an organism to be an effective competitor, can be viewed as a form of anticipation.
4. Complex adaptive systems have many niches. As change occurs, one niche is filled and another niche opens. In other words, new opportunities are being created by the system. The system is always unfolding and in transition. The key point about this last characteristic is that complex systems are not moving to some optimum configuration, but only changing to adapt to the other changes in the environment and the system.

I believe that this last point tells us an important thing about life on many levels. As individuals in a complex system, we can never hope to achieve some optimum level of success at which point there is no need for further improvement. Regardless of our level of success, there will always be someone to replace us. Our success is always dependent on our being a part of the system, and every part of the system, in some small way, contributes to our success. We, in turn, contribute to the success or demise of every other part of the system.

As a species, this point about complex systems should give us some humility. Humans tend to think of themselves as some ultimate end product of evolution that makes them masters of the earth, if not masters of the universe. In reality, we are only the best solution that nature has been able to produce through evolution to adapt to our current environment. Also, as a species, we are exceedingly dependent on every other living thing in our environment.

One of the problems that the human mind has with the theory of evolution, in spite of all the evidence to support the theory, is the fact that it is possible to conceive of a course of events where the human species might never have evolved at all. Also what does it mean for one species to be more advanced? Cockroaches have been around for several hundred million years longer than human beings, and they are good at being cockroaches. Microscopic parasites such as viruses are also good at what they do from a point of survival.

What we have learned about complex adaptive systems is that they are effectively open systems. In other words, we can never expect to come to some ultimate solution. Computer scientists have determined that in the game of chess there are about ten raised to the power of 120 possible combinations of moves, which is so vast that there is no way to provide an analogy. It can be said that ten raised to the power of 120 is larger than all the elemental particles in the known universe.

If the number of possible combinations in chess is so high, the number of possible combinations in complex adaptive systems is effectively infinite. The best we can hope for is to develop some rules of thumb that tell us what works best in a given situation. Continuing the analogy with chess, players do get better with time, and chess players are better today than they were fifty or one hundred years ago. Thus, we can expect to improve our rules of thumb even in an open system. The science of complex adaptive systems not only reminds us that "Life is a journey, not a destination," but it tells us that we can't ever know the destination.

Science may be able to make predictions in some cases, but in other cases, the best science can do is describe what we observe without ever being able to predict or tell why it occurs. However, we should go on trying to learn the why, because that is what makes us human. Some people might call this the religion or faith of science.

Adaptation in Complexity Is the Catalyst for Evolution

Adaptive problem solving explains how huge problems, such as the evolutionary development of the human brain, could take place from the starting point of basic amino acids in some primordial soup. The hierarchical building block structure of a system transforms a system's ability to learn, evolve, and adapt. Instead of moving through the immense number of possibilities step by step, an adaptive system can reshuffle building blocks and make giant leaps of improvement. An analogy might be that it is possible to locate a specific mailbox in the U.S. simply by using a nine-digit zip code. The first two digits narrow the search to a state, the next three to a section of the state, and the last four to a specific mailbox.

As powerful as computers have become in recent years, they still rely on the basic principle of binary code. In other words, all information can be stored and used as a series of zeros and ones, which could also be thought of as "on" or "off" switches in an electrical circuit. When we started to decode DNA a few years ago, it struck me as interesting that nature had evolved with a coding system that used one of four amino acids to create the sequences that we know as genes. Even more astounding to me was the elegance of the double helix, which could divide into two identical halves and then use the physics of molecular structure to simply and reliably replace the part that split away. When DNA splits, it reproduces itself in its entirety almost all at the same time. Compare this to the act of copying a computer file. We still have to copy one digit after another digit until the entire file is copied, and it takes some complicated technology to accomplish the process.

Co-evolution

Co-evolution involves symbiotic relationships. One such relationship is that between flowers that evolved to be fertilized by bees and bees that evolved to live off the nectar of flowers. Co-evolution can also involve an evolutionary arms race in which one species evolves to ward off another species, forcing that species to evolve into a better competitor or predator. A rabbit evolves to run faster to avoid the fox, and the fox gets faster at chasing the rabbit.

Co-evolution can also lead to cooperation in a species that has a memory. In computer simulations where cooperation is the best course, but cannot be verified, and only one cycle of competition is possible, self-interest prevails even though it does not lead to a favorable outcome. In other computer simulations where multi-cycle trials were possible and cooperation was advantageous, cooperation would take place if the cooperation could be verified and remembered.

The capacity of the human brain to remember and anticipate has resulted in social evolution. Social evolution increases our chances of survival as a species and provides benefits to our species as a whole. So far, social evolution has not reached the point where we have any impact on the genetic material that has brought our species to its position in the universe. However, we will soon be faced with issues that have no direct precedent in nature. If our intelligence is a result of evolution, can we use that intelligence to influence the process of evolution? Our generation or the next few generations to follow us will face issues where traditional values established through social evolution do not apply.

Social Evolution and the Future

The more we understand what it means to be alive, the closer we may some day be to creating something that, by our definition of life, is life. We are the first species that may have the potential to create our successors. Genetic research is already raising ethical questions. If we create something that is alive but does not behave the way we expected it to behave, do we have the right to destroy it? Will life become considered correctable, like a poorly built wall that must be reconstructed to keep it from some day falling on someone or causing the house to collapse?

Most people are willing to accept changes to genetic material that will prevent disease or premature death, but what about changes that just make our children more competitive in the marketplace? How will the rules have to change in a world that allows genetic tinkering?

Possibly the solution to this dilemma will be in social changes that no longer reward the individual for his or her achievements. This could not only go against biological and social evolution, but also against the political system in this country. In a system of scarcity, rewards for individual achievement have been the best way to achieve progress for the group.

The theory of the communist system was based on the concept that people received rewards based on their need, not their capability. However, this theory was never really put into practice. Also centralized planning proved to be a great failure in terms of getting the goods from people who could produce them to the people who needed them, especially when compared to the free marketplace. Thus, today we have no alternative social system that will work in a world where we have the ability to change human potential.

The Power Law of Change

The process of change in complex adaptive systems is different than change in simple systems. Change in complex systems follows the power law. When the power law is applied to earthquakes, the power law says that the power of an earthquake is in proportion to its frequency. In other words, there are many small earthquakes, but large powerful earthquakes occur infrequently.

As two of the earth's plates slide alongside each other, they do not move smoothly. Various geological structures create friction. The energy of the friction is stored by the distortion of the shape of the materials along a fault line. Eventually, just a slight bit more energy is stored in the fault than can be supported by the materials holding back movement. Once the fault starts to move, the fault reorganizes. In some cases, a few dislocations occur and the system then becomes stable. This is a small earthquake. In other cases, one dislocation cascades onto another section of the fault and so on until the cumulative energy of the movement is spent relieving the tension. This is what occurs in a large earthquake.

The power law also applies to evolution. Paleontologists, such as Stephen Jay Gould, say that we can see evidence of the power law working in evolution through the fossil records. We have experienced several periods of mass extinction. Scientists point to an asteroid as the reason for the last major shift in evolution, but it may be possible that a much smaller event could precipitate the next avalanche in evolution. Sometimes just one skier on a mountainside can start a huge avalanche.

In the political world, the events of September 11, 2001, are a perfect example of the power law. During the cold war, the tension between the U.S. and the Soviet Union created by mutually assured destruction (MAD) resulted in a long period of fixed, established structure for the entire world—not just between the two super powers. Since the fall of the Soviet Union, we have experienced a series of small wars in many parts of the world. In general, the world has become a more chaotic place over the last ten years. Many of these conflicts are releasing tensions that have built up over many years, not just the last decade. Since the conflicts have often been in less developed parts of the world, terrorism has been used to achieve political objectives. However, the 9/11 attack on the United States shocked the entire world. Suddenly, the world realized that terrorism could eventually endanger human civilization. The world is now moving to a new political order.

Probably, for the first time in the history of civilization, a major part of the world is unified in a call for political change. Naturally, there are two sides to the conflict. Americans may view our country's actions as an effort to eliminate terrorists. However, in the long term, the war will not be won until the issues that divide our global village are resolved to the satisfaction of mankind. The power law of change has cascaded to the highest level.

Perspectives

Every human alive today shares 99.9 percent of the same genetic material with every other member of our species. With so much in common with our fellow humans, one could expect that we all lead lives that are exceedingly similar and boring. However, if we look closely at nature, we see that randomness is the very essence of life. The life that every person lives is truly unique and, for the most part, unpredictable. We all have an interest in the future, but we really can't predict the future. In fact, most people really don't want to know everything.

For the most part, life is a journey and we do not have much control over what life sends our way. All life is complex, and through social evolution, humans have made it even more complex. However, it appears that nature is able to use this complexity to allow all forms of life to better adapt to the changes in the environment over time. The idea of working toward goals and objectives does not appear to be something that is inherent in nature. By the same token, since our lives can be influenced by so many factors beyond our control, we should be careful not to be consumed by working toward goals and objectives. Most important, we should not connect our self-worth with our achievements or our failures.

We should recognize that we live as a part of many complex organizations, and every member of the complex organization benefits by being a part of the organization. Everyone benefits from every small part that everyone else contributes to the whole.

While we may appear to be comfortable with stability and a lack of change, in reality, we are most comfortable with a manageable level of change. Life is a process of going through changes. From the time we are born to the day we die, we are constantly changing. Some people look at childhood and adolescence as the major periods of change in our lives because we are go through great changes in our bodies, but, in fact, we are constantly going through change, and the complex organizations that we are a part of also are going through change. While we may intuitively doubt our ability to live with change, science shows that as a species we are adept at living with change and adapting to change.

Charity is an attempt to share our good fortune with those who have been less fortunate. I believe that people are charitable because they realize that they enjoy benefits that were not entirely the result of their own efforts. Cynical people may say that benefits come at the expense of another person. While there is no denying that, in some cases, one person's gain is another person's loss, it is not the case most of the time.

Many people think that charity is the only way that one person can truly share his or her good fortune with another. In charity there is an obvious transfer of something of value from one group to another. However, as a society, we also share in our good fortune through co-evolution. When one person develops technology, it benefits many other people.

We should avoid thinking that good and bad events are the result of reward or punishment from a higher power such as God. We can be more successful by living life closer to the edge of chaos. If we expose ourselves to more things, people, and opportunities, we will be more successful and our lives will be more interesting as well. We will enjoy the journey more.

We have devised many social structures to protect us from the randomness of nature. We now have life insurance, auto insurance, and health insurance to protect us financially from the random events of nature. Mutual funds and banks protect us from random events in our investments and the ups and downs of our personal earnings abilities. We have been able to buffer our lives from the random events of nature, but we haven't been able to prevent the actual random events.

In this chapter, we have examined how nature operates every day in our short life times. It appears that nature has always operated the same way and will continue to operate the same way in the future. In the next chapter, we will look at our place in the universe in terms of time and space. We will examine how such an exceeding marvelous thing as the human body and mind could be created by nature.

Chapter Five

Time and Evolution: Our Place in the Universe

"Through no fault of our own, and by dint of no cosmic plan or conscious purpose, we have become, by the grace of a glorious evolutionary accident called intelligence, the stewards of life's continuity on earth. We have not asked for the role, but we cannot abjure it. We may not be suited to it, but here we are."

— Stephen Jay Gould

Stewards of the Earth

Understanding our place in the universe is an important part of finding meaning in our lives. It is important to understand both our place in the vastness of the universe and our place in the long history of the universe. An understanding can be at the same time humbling and satisfying. The dimensions of the universe are so large that they are practically incomprehensible in human terms. As far as we know, we are the only species on earth that has ever been self-aware and thought about such things as the meaning of life.

While we appear to hold a unique position on the planet earth, it is likely that we are not the only conscious beings that have ever existed in the universe. While we may never know for sure that conscious life exists elsewhere in the universe, the understanding that life could exist elsewhere provides a perspective on the meaning of our life.

We have learned a great deal about life on earth since the development of formal science about four hundred years ago. Today, through the use of communications and computers, our knowledge is expanding exponentially. In spite of all we know, we are just starting to realize all the things we do *not* know. Meanwhile, faced with these realities our desire to know more increases.

There are two exhibits at the New York Museum of Natural History that helped me with my understanding of my place in the universe. One exhibit shows the evolution of life on earth, or what some people refer to as the tree of life on earth; we'll look at that later in this chapter. The second exhibit, in the new Rose Planetarium, is a time line that shows the development of the universe from the big bang to current time. In the most basic sense, both exhibits are showing the same thing—a trend toward more complexity over the course of time. These two exhibits also deal with two areas of science where we have made some of the most significant breakthroughs in recent years: cosmology and evolution or genetics. Obviously, genetics deals with life. And, I believe cosmology deals with meaning.

Let's look at the time line that shows the development of the universe. While it provides much information about the birth and death of stars, the most impressive part of the exhibit is the one human hair at the end of the exhibit. The exhibit curves around for more than one hundred feet with the thickness of the human hair representing all of recorded human history. In other words, if you compare the life of the universe, which is approximately fifteen billion years, with the period of recorded history, it is equivalent to comparing a line over one hundred feet long with a line that is as long as the width of a human hair.

Paradoxically, the study of the very small—atoms and subatomic particles—through high energy particle physics is enabling science to develop significant insights into the development of most of the things that we observe in the universe. At the risk of over simplification, I will provide a brief description of the fifteen billion-year time line of the universe to give some perspective on the age and size of the universe and to provide some understanding of what the universe went through to get to the point where life started to develop on earth.

But first a word of caution. There have been many books on physics and cosmology in recent years that explain the findings of scientists without going into the mathematics and experimental details that are expected when scientists publish for scholarly journals. *A Brief History of Time* by Stephen Hawking is one of the more notable examples. Because these books exclude these details, they are sometimes said to be written for the layman. However, they are best understood by people with at least some training in science, because they build their scientific case on scientific principles developed by others. The whole purpose of science is to be consistent both with other science and observation.

To provide a description of the time line using this approach would take a book in itself. However, having read many of these books, I would like to observe that it is inspiring to see how science in the last one hundred years has been able to explain and provide consistency among various seemingly paradoxical phenomena. In effect, until recently, science has had two or more theories to explain the same thing. For example, the so-called big bang theory is now widely accepted by scientists as the description of the development of the universe. However, the term big bang was actually coined by a scientist who had an opposing view. He came up with the term as a way of showing his lack of respect for those who supported what can also be called an inflationary view of the universe.

The universe is trending toward more complexity. However, there is a principle in science called Okum's razor, which states that if there is a choice between a simple explanation and a complex explanation, the simple explanation is most likely correct. Science has shown that the most fundamental laws of nature are simple. For example, the formula most widely associated with Einstein's theory of relativity is $E = MC^2$. The universe took a long time to create the basic building blocks of life. While the universe is tending toward more complexity, there are some basic laws that drive these trends.

One other comment on the current state of science before I proceed: The explanation that I am about to provide deals with only four dimensions—the four dimensions that we can experience with our senses, which are height, width, length, and time. In the last ten years, physics has developed theories such as the super string theory and hyperspace that deal in as many as ten dimensions of reality. While it may take many years of additional work to fully develop the potential of these theories through empirical research and technology development, these theories include such possibilities as time travel and control over the forces of nature that are only science fiction today. Before you lift a skeptical brow, I would like to remind you that two hundred years ago harnessing the power of electricity and nuclear power were not even conceived of as science fiction. The main point here is that humans may soon have the ability to affect the forces of nature that we now think are only the province of some higher power.

The Speed of Light

The study of astronomy is unique among the sciences, because it is able to study things that happened millions or even billions of years ago as they actually happen. As we go about our daily lives, we see things going on around us via light reflected off objects as they move. As you are reading this page, light hits the page and is then reflected back to your eyes. Light travels at 186,000 miles per second. Thus, it appears that the information about an event that we get through eyesight is instantaneous. Anyone who has ever been in a thunderstorm knows that only when lightning strikes very close by does it appear that the sight and the sound reach us instantaneously. In most cases, we see the lightning and then some time later we hear the thunder. Since sound only travels at about 600 miles an hour, it takes much longer for the sound information to reach us. The event, the lightning strike, took place at a specific point in time. We became aware of the event by seeing the lightning, if we were outdoors and looking in the direction of the lightning strike. If we were indoors away from a window or outdoors and looking in another direction, we may not be aware of the event until another time when we hear the thunder.

If you went out at night when the moon was visible, you would be seeing the light from the sun reflecting off the surface of the moon. The light from the sun took about eight minutes to reach the surface of the moon, and it took a little over one second for the light to reflect down to you on earth. If you were to look at any of the stars at random, there is a good chance that you would be seeing light that left the star one hundred to a thousand years ago. With the aid of a telescope, you could see light from stars that has been traveling toward earth for one hundred thousand or a million years. In some cases, you may be seeing the light from stars that have already blown up or died thousands of years ago.

The night sky is like a motion picture history of the universe. The only difference between the night sky and a motion picture is that we are actually watching an event as it unfolds. For example, if we look at a star that is exactly 6,574 light years from earth and that star had a major flare-up that lasted for five days exactly 6,574 years ago, over the next five nights we would be able to watch the flare-up as it occurred over a five-day period 6,574 years ago.

An astronomer might point out that space is expanding, which would imply that it would take the light more than exactly 6,574 years to reach earth. However, the astronomer would also point out that over such a relatively short distance of 6,574 light years, the impact of expanding space would be minimal and may actually be canceled out by the gravitational forces of the large number of stars in our local portion of the universe. In any case, astronomy, like many other areas of science, is revealing aspects of nature that may not seem logical because they are not part of our day-to-day experience.

Every day the universe gets one day older. In terms of the universe, a year or even a thousand years is a short period of time. Therefore, even though events are continuously unfolding, we can look at different parts of the universe to see events that happened at a specific time in the past from one thousand years ago to ten billion years ago.

The Big Bang

In recent years, cosmologists have come to agree that the universe was born in an event that has come to be called the big bang. Cosmologists are mostly in agreement because they have been able to detect and measure the residual radiation from the first seconds of the big bang. In the fifteen billion years since the big bang, the background radiation has cooled to about three degrees Kelvin. Three degrees Kelvin is three degrees above absolute zero, or a point where matter has no heat or movement.

Approximately fifteen billion years ago, at the exact moment of the big bang, the universe was an infinitely small single point that contained a huge amount of energy. At this point, the laws of physics are unknown and possibly may never be known or are unknowable. Then the energy started to expand. As the energy expanded, time and space occurred or was created. This is referred to as the inflationary theory of time and space. The primary aspect of this theory is that space and time are created as the universe expands. For the average person, this is a difficult concept to understand, since our only experience is our movement about in space that already exists. Most people mistakenly assume that the universe is expanding in an empty void, but space is actually created, even today, as the universe expands.

In the first small fraction of a second, energy was so great that only energy existed. During this short time, all the forces of nature existed as one force, which allowed space time to expand rapidly to a size many times the size of the universe known today. As space time expanded, ripples in space time were created. As the energy expanded, it cooled down and the basic forces of nature were separated into strong and weak forces.

One of these forces, gravity, emerges and causes the expansion to slow down significantly. Also the energy cooled down until it was possible for some of the energy to coalesce or condense into matter to form the elemental particles such as electrons and quarks.

Still in a small fraction of the first second, some of the individual particles lost some of their energy, again as a part of the cooling process, and they were able to come together to form protons and neutrons.

At three minutes after the big bang, protons and neutrons were able to come together to form nuclei of atoms, since their binding energy was greater than the cosmic background radiation. The nuclei of basic light elements such as hydrogen, helium, and lithium were created.

At three hundred thousand years after the big bang, the universe cooled to about three thousand degrees Kelvin, and electrons could bind with nuclei to make neutral atoms. In a process that still goes on today, these hydrogen atoms came together or condensed into clouds of hydrogen. The clouds got larger and larger and through the force of gravity compressed into denser and denser clouds.

As this process continued, the gravitational force, which increases with mass, became so great that the hydrogen atoms started to collide with one another so forcefully that they started to come together to form new atoms that we call helium. Also in the process these atoms gave up some energy in the form of heat and light. This is the point when the stars started to light up.

By about the time the universe was one billion years old, this condensation of matter had formed stars, quasars, and proto galaxies. The ripples in space time created in a small fraction of the first second of the big bang provided sufficient irregularities in the distribution of matter to allow the condensation of matter into large astronomical structures such as galaxies. As we showed in the previous chapter, increased complexity comes from a chaotic region in nature. Thus, the creation of astronomical bodies also came from an area of irregularities or chaos.

This process increased the complexity of the universe again: It created new forms of matter as well as helium, and it altered the process of creation itself. The first stars were formed by a cooling process and expansion. Now in some parts of the universe there was cooling, while in other parts, the universe was heating up as stars formed and continued to "burn."

Creation of the Elements of Life

The process going on inside stars is the same process that goes on when we detonate a hydrogen bomb. Some of the mass of the bomb, or the star, is converted into energy. Only a small proportion of the mass is converted at any one time. Thus, stars can shine for billions of years. But they do eventually use up a significant amount of their mass. As a star loses its mass, the gravitational force holding the star together is overcome by the force of the nuclear reaction. Near the end of this process, the star becomes so large that it can no longer sustain the nuclear reaction. At that point, the gravitational force, with no other force to oppose it, causes the star to collapse with great force. These forces are so great that the hydrogen and helium atoms are forced together to form more complex atoms such as oxygen, nitrogen, and iron. These are the atoms that are required to form planets such as the earth and all the life on it. Once again, we have more complexity as stars burn out.

Thus, before life on earth could be possible, stars had to live and die. The atoms that make up our bodies come from the stars that existed billions of years ago. Although our lives are so short in relation to the age of the universe, you and I exist because all that went on in the billions of years before. And today we are sustained by one star: the sun. If we suddenly lost the heat of the sun, all life would stop in a matter of weeks.

Five billion years ago, when the universe was already about ten billion years old, our solar system was formed from the remnants of earlier stars. We are relative latecomers to the universe. After expanding for fifteen billion years and creating new stars continuously, it is hard to comprehend the number of stars there are in the universe. Our galaxy, the Milky Way, which is about ten billion years old, contains approximately two hundred billion stars. There are billions of galaxies in the universe. In order to try to put the number of stars in the universe into perspective, think of the number of grains of sand on all the beaches and desserts on earth. There are more stars in the universe than there are grains of sand on earth.

Through direct observation, science has shown that the laws of physics work the same throughout the universe. Intuitively, we could also come to this conclusion, since the big bang theory says that the entire observable universe came from the same origin fifteen billion years ago. Thus, we could also conclude that the same laws of nature that lead to development of life on earth would also lead to the development of life in other parts of the universe. I believe that we can put aside the question of the existence of God in this scenario, since the laws of nature that resulted in the development of life would still be consistent throughout the universe whether or not they are guided by the hand of God. A theologian might say that since consistency is more eloquent than chaos, the existence of God could only reinforce the probability of consistent laws of nature throughout the universe. However, as we saw in the previous chapter, chaos may be more eloquent than logic would indicate.

We have a good idea of the conditions that were present on earth when life developed. While we would intuitively conclude that other stars in the universe would have planets revolving around them, recent observations with the Hubbell Telescope provide data that indicates that other stars do, in fact, have planets revolving around them. Naturally, all planets do not provide the conditions that allow life to develop and evolve. In our solar system, it appears that there is no life on the other planets, although the possibility exists that life may have existed on Mars at one time.

However, assuming that only one in a million planets provide the conditions that could allow the development of life, there would be around two hundred thousand opportunities for life to develop in our galaxy alone. Some people would ask: if there is such a great potential for life, why haven't we heard from them? While stories of UFOs are common in the mass media, there is no hard evidence that most scientists would accept that even hints at the existence of extraterrestrial life. In addition, the SETI (Search for Extraterrestrial Intelligence) project has been looking for life within one hundred light years of earth for many years without results.

There are two factors that make it easy to assume that we are alone in the universe. Those factors are time and distance.

Large Numbers

We have become accustomed to hearing and talking about large numbers. The gross domestic product is expressed in terms of trillions of dollars, the sales of major corporations are reported in terms of billions of dollars, and the population of the United States is in the hundreds of millions. When scientists say that the universe is fifteen million years old, it is difficult to comprehend how long that time really is. Carl Sagan had a way of making these time frames more meaningful. We have a good sense of a year of time and the way we break it up into months, days, hours, minutes, and seconds. Carl Sagan asked us to imagine that the universe is one year old and then compress other time frames in proportion.

For example, the earth is about four billion years old. So in terms of the life of the universe, if the big bang occurred on January 1, the earth was formed about September 27. If humans or our direct ancestors have been on the earth for about four million years, humans started walking the earth on December 31 at about 9:40 in the evening. We have some fragmented records, not the fossil records, about man's activities on earth that go back about five thousand years. In our new universal time of one year, the last five thousand years are equal to the last ten seconds of the year. The ball in Times Square is almost all the way down the pole. If you are now eighty years old, your time on earth is less than two-tenths of a second on the clock.

The universe is so large that astronomers measure distances in light years. A light year is the distance that light travels in one year. Light travels at 186,000 miles per second. It takes the light from the sun about eight minutes to reach the earth. The next star nearest to the earth is a few light years away. Astronomers use a light year as a way of measuring distance because it is a convenient way of expressing the large distances that exist between objects in the universe. But a light year is also a good measure because it is a convenient measure of communications between different points in the universe. Light and electromagnetic radiation such as radio waves travel at the speed of light. Since Einstein's theory of relativity states that nothing can travel faster than the speed of light, the speed of light represents the upper limit of the speed of communications between different parts of the universe.

Electromagnetic Radiation: Is Anyone Out There?

There are two basic ways that we observe objects in the universe: either through the light that they emit or, in a few cases, through the light that they reflect (such as the moon and the planets in our solar system and radio waves that they emit). Stars, including our sun, emit electromagnetic radiation as a part of the process of converting mass into energy.

If we want to look for extraterrestrial life, we can either use optical telescopes or radio telescopes. Optical telescopes aren't good for looking for life even in our own solar system, because the signs of life are not easy to see over distances on the scale of our own solar system. All the life we see on earth and all the geological structures, such as mountains, occupy just a thin film on the surface of our planet. All life exists within a layer that is less than six miles thick. From space, the earth looks smoother than a cue ball. Thus, radio telescopes are about the only practical way to look for life in other parts of the universe.

But before we can detect life by looking for electromagnetic radiation, the life on a distant planet must have evolved to the point where it has been able to use electromagnetic radiation as a means of communications. Although life has roamed the surface of the earth for hundreds of millions of years, and humans, which we classify as intelligent life, have walked the earth for four million years, we have used electromagnetic radiation for communications for only about one hundred years. If some other advanced civilization in another part of the Milky Way is looking for intelligent life, they will not be able to detect us unless they are within one hundred light years of earth.

Although one hundred light years may seem a long way, even to someone who can get in a plane and fly from one coast of the United States to the other coast in five hours, one hundred light years is a short distance on the scale of even our galaxy. Going back to our universal clock where the life of the universe is one year, our "on the air" light has been lit for about three tenths of a second.

The Rate of Change in Technology

There are many factors that affect the rate of change in technology or in a broader sense human progress, but the most important factor is the size of the group that is working on solving the problem. Archeologists tell use that the human species started as hunter gathers, which is not much different than other animals. As hunter gathers, we lived in small groups that moved around following the source of food. When we started agriculture, we stayed in one place, which allowed us to live in larger groups, with some division of labor or specialization. The larger groups also provided us some protection from the risk of competition from other groups. Eventually, we learned that trading or commerce among groups allowed us to improve our standard of living. At this point, the social group we participated in could involve people from a geographical region.

As we have worked in ever larger groups, we have improved our communications capabilities. Animals, and maybe even humans in their early development, communicate on an instinctual level—in other words, through habits that are learned and then passed down through genetic information. Some scientists say that the development of language is the most important factor for human development. In any case, humans developed verbal communications and then written communications. Today we use verbal and written communications assisted by electromagnetic energy, which we call information technology, but we have really been working with information technology from the time that we started talking to each other. Improved communications now allows humans to participate in a global economy.

Maybe one of the reasons that the term information technology was coined is the fact that we noticed how fast our communications technology was changing. The technology has been changing for thousands of years, but until recently, the change in any one generation was not noticeable. The main point is communications and the other technologies that communications allows changes exponentially.

We only have our own history to use as a guide, but it appears that once technology change moves into its high growth phase, it changes much more rapidly than the changes in the social organization of the society that is developing it. We know from history that humans have always been in competition with each other. Competition is part of evolution. However, we cannot say that competition alone is the main theme of survival. It appears that cooperation is also part of survival, either as a learned behavior or as some part of basic nature as we have shown with self-organizing complex systems in the previous chapter.

The problem appears that the exponential growth of technology eventually puts any intelligent species in a period of time when it has the capacity to use the technology for its own self-destruction. This self-destruction may even be unintentional. Many people point to the dangers of nuclear energy. Humans' first use of nuclear energy was to create a bomb, which it used at least twice to kill humans. We have since used nuclear energy for peaceful purposes, but we are today contending with the problems of disposing of the waste of nuclear energy—the spent fuel rods. Eventually, we will solve the problems with fusion energy and will be able to use nuclear energy for the benefit of mankind.

A less obvious and thus less intentional self-destructive use of technology is all the uses of fossil fuel to create energy. Although most people are convinced that the greenhouse effect is real, we are less certain about how serious the problem is and how best to address the problem.

The least obvious but potentially the most self-destructive use of technology may be in the area of biology and medicine. Putting aside the military use of medical technology, some of the technology that we use to cure disease and prolong life may eventually lead to our destruction. We may learn some day, when it is too late, that the antibiotics that we used to cure infection will over the long term leave us vulnerable to extinction as a species as a result of the emergence of an infection that is resistant to all antibiotics.

I have provided all these examples, not to make a case against technology, but rather to show that once a civilization reaches the point of being able to produce electromagnetic communications, it is also in a period, when it may self-destruct either through social immaturity or by an unavoidable technological accident.

It appears that intelligent life passes through a period of hundreds or maybe thousands of years when the advances in technology that allow communication also allow self-destruction. In addition, it appears that there are other events that could result in the extinction or a severe setback of intelligent life before it can be detected by another civilization. Some of these events include massive volcanic eruptions, major climate changes such as an ice age, and impacts of asteroids or comets. The asteroid that impacted the earth and killed the dinosaurs happened sixty-five million years ago. The last ice age ended less than ten thousand years ago.

Other Civilizations: Odds Are We're Not Alone

There is a good chance that we are not the only life that has ever existed in the universe. Consider the following scenario, which illustrates how other life can or could have existed in the universe and why we may never be aware of its existence, even with more sophisticated technology than we have today.

Here are the ground rules of this scenario: Time is passing on our universal clock at the rate of one year for every fifteen billion years. We have suspended one law of physics just for ourselves—we can see the expansion of electromagnetic radiation instantaneously rather than waiting for it to reach us at the rate of 186,000 miles per second. Also this electromagnetic radiation is visible to us as blue light.

So here we are sitting and watching our galaxy. It is the first day of December. Way over in the left hand side of the night sky we see a blue light go on. Slowly it expands from a point to a small dot to a shape about the size of the moon. We then notice that the blue light in the center of the circle has gone off. Maybe it was on for thirty seconds in total. The circle continues to expand like a ripple on a pond. As we continue to watch, we see another blue light appear in the middle of the sky in front of the ripple; it stays on for just a few seconds before it goes out. Almost at the same time as the second blue light went on, a third blue light appeared inside the expanding ripple. Light number three stays on and a fourth blue light appears, also inside the ripple, at a distance of about the diameter of the moon from the third blue light. Light number four also stays on.

This little scenario shows what might actually be happening at this very minute in the Milk Way. Let us assume that the fourth blue light was the earth. Let us go on to speculate what might have happened with the four civilizations in our scenario.

The first civilization (blue light number one) was sending out electromagnetic radiation for fifteen thousand years. It had taken almost five hundred years after it discovered how to use radio before it learned how to use nuclear energy. This was due to the fact that this civilization was a relatively peaceful society with a large number of nation states that were able to work to mutual advantage. It was also a civilization dominated by one religion. The civilization eventually did develop nuclear energy in the form of an explosive, which it stockpiled and maintained in the event that it had to fragment an asteroid that was headed toward its planet. It never had an opportunity to use these explosives. This civilization ended on the day that some physicists were conducting experiments with anti-matter. Due to the effect of some yet undiscovered force in nature, the experiment created a black hole that consumed its entire solar system of twenty-two planets plus four other solar systems in the immediate vicinity. On one of the other planets in these solar systems, life had developed to simple multicell organisms.

The second blue light represents a civilization that learned how to use electromagnetic radiation and nuclear energy in the same century. However, this planet experienced a massive volcanic eruption that eventually resulted in the death of 96 percent of the planet's population. The remaining 4 percent of the population survived in remote areas where they were still living in Stone Age conditions. When the ripple of electromagnetic radiation from the first civilization passes over this planet, it will not be noticed. Civilization number two has yet to restore its infrastructure to the point where it is able to detect that civilization number one ever existed.

The third and fourth blue lights appeared after the electromagnetic ripple from civilization number one passed over them. Since electromagnetic waves leave no lasting imprint that can be read at a later time, civilizations three and four also have no way of knowing that civilization one ever existed.

Both civilization number three and number four—the earth—learn about each other's existence about the same time. Since it took fifteen thousand years for the electromagnetic ripple from planet three to reach the earth, the earth now is much more technologically advanced than when the earth's electromagnetic ripple first started passing over planet three. The earth is now governed by a planetary government that shortly after the excitement of learning about civilization three adopts a resolution to monitor the transmissions from civilization three to learn as much as possible about this civilization. The two areas of interest are: technology that may not have yet been developed by earth, although this may be unlikely given the fifteen thousand-year time lag; and the possibility of a threat from planet three at some time in the future.

Although I have mixed some speculation with science in the previous scenario, the example is based on the latest science. While most people may think of the human species as the only instance of intelligent life of all time, there may have been, in fact, hundreds or thousands of civilizations that existed throughout the universe over the last five billion years. Also, while in terms of the human experience, life seems permanent, life may be quite transitory in terms of the age of the universe. When we think about eternal life, it may be helpful to think of the age of the universe to date to put the term eternal life into perspective.

Some people may ask what does all this have to do with a personal meaning of life. I believe that we all create meaning from what we know and what we personally experience in life. Thus, this information will have a different meaning for each person. It may or may not reinforce your belief in a higher power that some people refer to as God. In short, belief in a divine being is still a personal, and not a scientific, decision.

In my mind, if I compare what we know about the universe today with what we knew in the time of Galileo and Copernicus, we are much more certain of the physical world in terms of the natural order. Some people may say that the time and the distances discussed above are so great that they have no practical relation to the time and distances we experience on a day-to-day basis, but it does provide some point of reference.

Cosmology to Evolution

Up until this point, creation, if you want to call it that, is taking place at the level of astrophysics. Also, up until this point, the trend toward complexity is a relatively slow process. To create basic atomic particles and all the hundred or so atoms that appear naturally in nature took several billion years. It probably took half of the time span from the big bang to create this level of complexity. It also means that the atoms that are a part of our bodies were at one time a part of and came from the stars. Thus, we may think about our lives as conscious beings as being short, two-tenths of a second in our earlier example, when in reality the atoms in our bodies are as old as the beginning of creation.

After the earth and other planets were formed, either in our solar system or around other stars in the universe, the march toward greater complexity picked up pace. In recent years, astronomers have been able to detect the existence of planets around other stars. Given the immense size of the universe, it is almost a certainty that another planet with the same environmental conditions as earth exists.

Evolution Equals Adaptation

Scientists tell us the process of evolution is simply the net result of reproduction, random changes in genetic material, and natural selection or "survival of the fittest." These changes do not occur often, but over long time frames these changes do occur enough to result in new species. As these species develop, some species are not adept at living in their environment. Thus, the law of the survival of the fittest determines which species will prevail. It could be said that the law of the survival of the fittest is relative. In other words, the survivor is not the strongest species but the species that works best in its environment.

This process works when changes in the environment are slow because the process that causes the change is also slow. In recent years, the theory has developed that the dinosaurs died off due to a random event—a large asteroid striking the earth—that caused a rapid change in the environment. These animals could not live when the sun was blocked out rapidly by the dust and smoke that resulted from the asteroid strike. Smaller animals, more adept at surviving under the new conditions, prevailed.

An amazing development in nature has been the ability to record information in the form of genetic material. In fact, without this recording system it would not be possible for evolution to take place. There would be no communications system to pass the net result of all the generations that came before. This is even more vital when the number of generations between evolutionary changes is great.

It is interesting that humans developed a coding system for processing information in digital computers that is quite similar to the coding system used by DNA in genetic material—and this coding system was developed before we understood the makeup of DNA.

The coding system for computers uses sequences of zeros and ones to store and process information. These sequences are called bits and bytes. It takes eight bits (either a zero or a one) to make a byte. Eight zeros or ones can be combined in 256 unique combinations. Thus, eight bits to a byte allows computers to have 256 characters to store and process information. In the English language, we use twenty-six letters and ten digits to store information on paper, which we now refer to as hard copy. Computers use zeros and ones to store information because computers are electrical devices and an electrical circuit can be either "on" or "off."

DNA uses sequences of four amino acids to store information. Since the building blocks of life are molecules, nature uses molecules to store information.

The one major difference between computer code and DNA is that humans have used computer code to store and process information after nature has evolved. Computer code can be changed by programmers, as computers evolve. It is a system that humans have imposed on top of nature. DNA is a system that evolved with nature and developed or became more complex step by step as nature evolved and became more complex.

Also both systems use simple material to store information. The genetic code is so good that it does not just record the characteristics of a species but the code for a specific individual. This system is so complex that only with the aid of computers are we now able to start reading this code.

We have just recently been able to map out the entire human genome. It is an awe-inspiring thought that something so complex as the DNA molecule, the blue print that determines everything about us, is carried by every one of our cells. However, when you understand that DNA was the starting point of life, rather than something that was added later, then you also can understand that this was a part of the natural process toward more complexity. We sometimes refer to DNA as instructions, a blueprint, or a program in order to provide an analogy with the things we understand. When humans build something, they create a blueprint of the finished product, or they write an instruction book after the thing is finished. But DNA changed as we evolved.

In the early days, as the earth cooled, it was covered with a chemical soup. The basic atoms that make up life were reacting with one another forming more and more complex molecules. This process continued until a unique molecule formed which was the precursor of the molecule that we know today as DNA.

This molecule, which is in the form of a double helix, has the ability to duplicate itself. When this molecule splits into two separate pieces, the two pieces have the ability to attach the molecules that were lost in the split with other free smaller building blocks to create two exact copies of the original DNA molecule. These molecules then developed into cells.

Through early stages of evolution, these cells started to develop special functions or capabilities. Thus cells came together to form more complex forms of life. Although these forms of life are simple in comparison to what we think of as life, the process continued over hundreds of millions of years. It was only about 500 million years ago that life started to move out of the sea and onto land.

The forces that are part of the natural selection process can range from other forms of life such as animals, bugs, or even viruses that result in the demise of a species to changes in climate to catastrophic events. Dutch Elm disease can kill off an entire species of trees, as can tent caterpillars. Humans almost killed all the buffalo. The dinosaurs died when a medium-sized asteroid slammed into the earth sixty-five million years ago.

Dinosaurs roamed or ruled the earth for about one hundred million years before the impact of the asteroid. Mammals, our branch of the tree of evolution, lived in the time of the dinosaurs, but they were not able to progress beyond the level of small rodents. The rodents managed to live through the changed environmental conditions that existed after the impact, while the dinosaurs could not. Without the competition from other larger animals, evolution allowed mammals to eventually rule the earth. Thus, it is reasonable to say that without the impact of the asteroid, a random event that occurred sixty-five million years ago, the human race would not exist today.

We Are a Part of Nature

The other exhibit in the New York Museum of Natural History that I mentioned earlier shows the evolution of life on earth. It could be easy to miss the connection between the evolution of the universe and the evolution of life on earth, but they are both part of a larger process. The most striking part of this exhibit is subtle but, I think, more profound than in the first exhibit. The exhibit shows man at the top of our branch of the evolutionary chart. Our branch (Homo sapiens) started about four million years ago.

If you study the chart, you see that life is continuous—each species and generation within the species supports the one that came before. Also change is almost continuous, and genetics tells us that change is the result of this continuous process. At some point, a subtle change occurs in a species and a new species is formed. Unless some catastrophic event occurs, such as a huge asteroid hitting the earth, we as a species will evolve also. One thing is certain: We are all links in the chain. We all form a tiny part of the time line, but it is essential that we all do our part.

As humans, we think of ourselves as somehow different than everything that came before us. Since we are self-aware, some think that we have a spirit or soul that is unique to humans. However, if we look at some of the traits that we think make us unique, we find these traits in non-humans as well. Other species can communicate among themselves, are capable of social interaction, and can learn (although some may call it conditioning). However, there is one thing that makes humans unique. We have been able to take learning far enough to discover the fundamental laws of nature. We have also developed the capability to end the chain of life established on earth either intentionally or by mistake. All of the changes that have occurred over billions of years can be attributed to random events. At least, we can explain how one event led to another without the need to include some unexplained factor. By the same token, we are the first species that we know of that has the capacity, or will soon have the capacity, to change the course of evolution. Or it maybe said that we are the first self-intra-acting species in terms of evolution.

Although as a species we have been able to reach this point in the universe, the major part of our life is still consumed by the essentials of life—we still have to eat, sleep, reproduce, raise our young, and care for the sick. Whether we work on an assembly line or as a CEO, the car we drive, the house we live in, and the person we have sex with are all just a matter of scale or choice. Even the few of us who work to discover the fundamental truths of nature are still driven by basic biological needs. By the same token, we can all equally enjoy music, the World Series, or a beautiful sunset. Most important, if we didn't work together, we would not have been able to know everything we know today, and enjoy the benefits of this knowledge.

Some might say that this analysis does a good job of bringing us all down to some common denominator. I would rather say that it makes us all a part of a universe that has evolved from a single point of intense energy to a complex system that has been in existence for billions of years. The planet earth and everything on it is so small in relation to the rest of the universe that it fades almost to nothing. But each of us is more than the entire earth. We are made of star stuff.

In U.S. society, we place great value on individuals and their freedom to make the most of their lives. This analysis may seem to diminish the value of the individual and minimize the struggle we all endure to make a better life for ourselves and our children. Rather, I believe that when we put that struggle in perspective with who we are in relation to the universe, it can make some of everyday concerns less of a burden. Also it levels out some of the highs as well as the lows. In answer to the question: *Why am I here and what is the purpose of life?*

Our purpose is to make whatever contribution we can to improve the overall quality of life on earth for both humans and non-humans. In cosmic terms, human life is short, but our contribution to the knowledge base is everlasting.

First Generation to Manipulate Genetic Material

We are the first species and among the first generations of our species that will be able to change genetic material by manipulating it rather than waiting for random events to occur. In a sense, we will be using the approach we use with computers. If we want to change the way the computer operates, we change the computer's program. If we want to change how a human turns out, in the future we may change the genetic program.

We could also use the analogy to make a distinction between how our minds work and the way computers work. When we see something, that event is recorded in our brains by something changing in our brains. The network of the neurons in our brain changes. When a computer processes information, it goes through a series of steps that have been established by the design of the computer hardware and software.

To develop artificial intelligence, which means that a computer learns from its information input, a computer will have to make changes to its system. As computer systems become more complex, it will eventually become difficult to distinguish between what it means for a human to be alive and when it means for a computer to be alive.

From this time forward, we will be able to write history at least in terms of evolution. By changing the genetic material, we are, in a sense, creating the outcome of the random event in history rather than waiting for it to occur and then be recorded. I am sure that is why people are so cautious about genetically engineered crops and cloning. We may create a Frankenstein before we realize what we are doing. This also means that we will be able to greatly accelerate the process as well as determine its direction. If we can use our experience with electronic computers over that last fifty years as a model, this change will be rapid indeed. It could be said that this is the first major change in the way that humans interact with their environment since men first started using tools to augment the capabilities of their bodies.

Understanding our DNA will eventually allow us to be a part of the selection process. Some people are concerned about the possibilities that this provides. But humans already use science to affect the natural selection process even if it is in a more subtle way. Today we use it to cure disease; later we may use it to improve our species. However, I believe that it's one thing to cure disease and another thing to make us more intelligent or in some other way better competitors in the survival of the fittest.

Evolution and the Individual

When we talk about evolution and survival of the fittest, most people think in terms of the human species or Homo sapiens, if we narrow it down a little. However, I believe that an understanding of what we are learning about this process can help each of us understand our place in the universe. All humans share the same basic genes, and humans share a high percentage of the same genes with other species that are close to us on the evolutionary tree of life. On the other hand, everyone's exact genetic makeup is different from everyone else's. This fact is one of the reasons that gene testing has been used in recent years in criminal investigations to make positive identification of suspects.

Each of us carries around in our bodies a complete genealogical record of all the generations that came before us. This record contains about three billion characters. Contained in the record is information that goes back millions of years, and this is the information that we share with many other members of our species. The information stored in our genes from the more recent past is shared by fewer people, and these are the genes that make us increasingly more unique as we approach our generation.

This information not only provides a historical record, but it also determines if we will be susceptible to certain diseases such as tuberculosis. One of the most significant implications of the knowledge that we all have unique genetic makeup is that we all have a unique potential.

In the United States, we recognize the uniqueness of the individual while at the same time saying that all men are created equal. I believe that the spirit of this statement is true, and it may also be true in fact at some level. However, when we get down to the level that science is capable of taking us today, some are more fortunate than others, in terms of what they have inherited from their ancestors.

I would prefer to look at our inherited genetics not as a bank account but rather as pages in a history book to which we have the opportunity to add additional pages. More important, in terms of thinking about the meaning of life, it shows that we are all beneficiaries of the many years of evolution, for which we cannot take any responsibility or credit. I have said more than once in this book that life is a journey rather than a destination. I believe that this statement could be modified to say that our individual lives are a continuation of the journey started millions of years before our birth.

What we are learning today about genetics raises many issues. Is it ethical to modify a person's genetic makeup to provide him or her with traits that will most likely lead to more successful lives, at least in terms of what most people consider success? We have already seen how the assessment of potential impacts opportunity. SAT scores are already used as a means of determining a student's potential for learning. Generally, the more desirable—and academically exacting— colleges require higher SAT scores.

Our Place in Nature

We can talk about the vastness of the universe and how long it has existed, and we may be humbled or inspired by the thought that we are part of such a universe, but we also know that all the life we personally will ever know will be the life we live on the planet earth. Thus, knowing our place in the universe becomes more practical when we look at our place in the evolution of life on earth. One could say that knowledge about anything else other than life on earth really has not impacted our life on earth, with the possible exception of people such as Copernicus and Galileo whose theories of the universe created controversies that changed their lives.

It is a fact of life that the human species dominates life on earth. As a part of the human species, our life is impacted continuously by the fact that we hold this dominant position. In countries such as the United States, every member of this elite group is assumed to have certain inalienable rights simply because we are members of the human species. The Bible also teaches that God created the earth and gave man dominion over it.

Religion teaches us that what makes man different from the rest of nature is that man has a soul, which is some spiritual form that is combined with the body while it is alive and which is separated from the body at death. Science tells us that consciousness or intelligence is the one feature that makes man different from the rest of nature. Science tells us that our capacity to do the things that we define as a part of intelligence such as our ability to use language and to plan ahead are possible because of our large brain size. The difference between religion and science could be summed up by saying, that religion believes that what sets us apart from the rest of nature—a soul—is not a part of nature, while science says that while we are the dominant species on earth, we are still totally a part of nature.

Part of our inclination to believe in the supernatural is due to a belief that whatever cannot be explained must be due to some supernatural forces. Consciousness and the working of the human brain are still in the realm of things that are so complex that it is difficult to even anticipate when we might have some understanding of how the brain works. However, our ability to understand other things and to identify and explain other forces that we did not know existed less than a hundred years ago, opens the possibility that we might at some time be able to explain the working of the brain and consciousness.

Social Evolution

Some people maintain that one of the shortcomings of the theory of evolution is that it cannot show that nature has any purpose beyond survival through adaptation to the changing environment. The problem with this analysis is that there is the tendency to confuse biological evolution with social evolution. Most of the progress that humans have made as a species has been during a time in which there has been no change in our basic genetic makeup. While we have seen some improvement in our physical and intellectual performance, we have not experienced any significant physiological changes since our species emerged about four million years ago. Social evolution can be considered a part of evolution, since the main purpose or benefit that has been realized by social evolution has been the increased survival capability of the human species.

Our basic intelligence was a product of evolution, but our ability to dominate our environment has been a result of our social evolution, which was significantly enhanced by our intelligence. Other species are able to work together as social creatures, but these capabilities are the result of habits that are passed on through the genetic record rather than any capacity to anticipate and plan for future contingencies.

There are two characteristics of social evolution that have allowed the human species to accelerate its development: Humans can transfer directly what they have learned to their young by teaching them what we have learned. In biological evolution, nature must wait for physical mutation to emerge that is more adept for the current environment and then allow the offspring with this mutation to dominate in subsequent generations.

Just as humans can pass along what they have learned to their young, they can also share their invention with other social groups. Thus, the rate of invention can be accelerated by the size of the society that does the inventing and the resources of that society that can be devoted to innovation. The human species is the first species that will leave something in the geological record other than their own bones.

On the darker side some scientists believe that we are not evolving fast enough socially to protect us from the improper use of technology. It is possible the improper use of technology will ultimately result in the extinction of our species. Thus, we will not be the ultimate species to dominate the earth, which also brings into question the purpose of evolution especially with regard to the human species.

Beside the fact that we are designed to survive and pass our genes on to the next generation, science has not been able to tell us much about the purpose of life. It appears that nature really doesn't design with a purpose. Often the species that survives has redundant capabilities. The redundant capabilities are completely redundant except in the case where one capability allows the species to survive when other species do not have the redundant capability.

For example, at one time some fish had both gills and lungs. It was this capability that allowed some fish with other characteristics suited to life on land to move out of the sea and onto land. The study of evolution leads us to the conclusion that the purpose of evolution was the human species. In fact, the laws of probability say that if we could run the sequence of events again, the human species may never have evolved.

In a scenario where an asteroid did not hit the earth sixty-five million years ago, the dinosaurs would still be alive today, and mammals would be small creatures living in the ground. Before they went suddenly extinct, dinosaurs had dominated the earth for one hundred million years. There is also little chance that dinosaurs would have evolved to have large brains. They had been successful for one hundred million years without the benefit of large brains.

On the other hand, just because life has no purpose, it does not automatically follow that life has no meaning. Life is open to us, and we have potential, which provides meaning. Most people seem to accept the fact that life has limits. Thus, why should the fact that we are mortal lead to total despair? In fact, as we get older, we accept the fact that we lived life in stages and that at some point our purpose in this life is completed. We have children and grandchildren. Even if we had an important job or made a major contribution to the arts or science, there is only so much one person can do.

Outside of man there does not appear to be any behavior in nature that shows morality. In fact, there are many things in nature such as the predator-prey food chain and the parasitic behavior of some organisms started at the microscopic level on up to human-size animals that are completely abhorrent to human sensibilities.

Unfortunately, it is almost impossible for any scientists to approach the study of any area of science without some biases that are so ingrained that it is hard to recognize them as biases. Western culture has a bias toward the belief that society is moving toward some form of culmination, either good or bad.

Scientists may agree on the explanation of many things we find in nature, but scientists still disagree about the need for an explanation for all things in nature. Some will say that understanding increases the wonder, while some others say that understanding takes away the wonder.

Personal Meaning

Up to this point we have looked at what we have learned about nature or what *is*, in an effort to determine if we can find some absolute meaning of life that every person can use as a reference in finding meaning in his or her own life. I hope I have shown that, in fact, there is order in nature. There is an order that appears to unfold over time, but paradoxically it also appears that the order works in an environment of uncertainty or randomness. The complexity created by random events makes it impossible to determine if nature has any direction or goal, other than to maintain some sense of order as nature evolves.

There are many laws in nature, but only some of those laws provide direction with regard to laws for social behavior. For example, nature appears to make progress through an interaction of many forces—some working together (in a complementary sense rather than a cooperative sense) and some working against each other. I believe that the success of U.S. society is due to the fact that our social organization seems to be more in alignment with nature than most other societies that came before ours.

While nature is vast and every person is the summation of billions of years of nature's changes, every individual possesses a great number of capacities to interact with his or her environment without the aid of any other part of nature. We are not all cogs in some great machine. We are individuals with free will, mobility, the ability to communicate, and, most of all, a separate conscious experience. On the other hand, we also depend on nature for our survival, and we owe our very existence to nature. To borrow a phrase from the Declaration of Independence, "These truths are self-evident." No faith is required in someone who has told us that these things are true.

Since every person has his or her own conscious experience and since every person has a free will, every person also has to establish his or her own personal meaning of life. In the remaining chapters of this book, we look at how individuals establish meaning in their life. Unfortunately, many people experience a great deal of frustration in their lives. As one person once said, they live lives of quiet desperation. The goal of this book is to help you evaluate and enhance the process of finding meaning in your life and establish a genuine sense of self-worth.

Chapter Six

Personal Meaning:
Purpose, Self-Worth, and Security

"I agree with the guy who said that I am all for progress, it's the change I don't like."
—Mark Twain

There's Self – And Everything Else

Philosophy and science have pondered individual experience since we were able to wonder about it. The more we look at consciousness, the harder it is to explain. It is so wondrous that it is possible to assume that it is not even a part of our physical bodies. It seems to take place in our heads, but we don't know where. Fortunately, for the purpose of this book we all have a direct experience with consciousness and that should be enough to provide a reference for what is presented in this chapter. We all have an experience that we call our *self.* There is the self—and everything else. It makes no difference how large the universe is or how long it has existed; it is everything else. We are a part of the universe, but at least in some sense, we experience it as being outside of ourselves.

It is said that life defies the second law of thermodynamics. The essence of the second law of thermodynamics is that the universe tends toward disorder or that it is winding down like a clock. If we wait long enough, probably more than one hundred billion years, the universe will eventually be completely devoid of everything including heat. We experience this law every day of our lives. Metal rusts, water gets dirty, our coffee gets cold after we take it out of the pot, things fall apart, and living things die and decompose.

Life—no matter how simple or complex—is able to take resources from its environment and use those resources to work against the second law of thermodynamics. Plants can take in sunlight and carbon dioxide and make sugars and oxygen. Animals can take in plants and oxygen and make proteins. Thus, in the most basic sense, we are constantly in a battle with the second law of thermodynamics. Once we lose the battle, we are dead, and as far as we know, so is the end of our conscious experience. In fact, our conscious experience can end even before our bodies are dead. People live in comas in hospital beds, with no higher brain activity, for years.

According to Abraham Maslow, humans are motivated by a hierarchy of needs. Maslow says we focus on our most basic needs first. Then only after those needs are satisfied, do we focus on the next higher need, although it is possible to satisfy two levels of need at the same time. For example, if we are deprived of air, we will do everything in our power to get air. In the process of determining if a person is brain dead, doctors will turn off a respirator and wait to see if the patient will start breathing on his or her own. Breathing is such a basic function that it is assumed that we will breathe on our own if there is any activity whatsoever in the brain. Likewise, moving up the ladder, studies have shown that when people are living in conditions of starvation, they lose all interest in sex. Thus, while food and sex can play a role in the meaning of life, any consideration of what makes a life meaningful only comes into play after the basics of survival of the species have been satisfied. The needs start with physiological needs and move up through needs for safety, love, esteem, and self-actualization. Self-actualization involves realizing one's potential.

Our daily battle with the second law of thermodynamics goes beyond our taking in sufficient resources to maintain our bodies, and the human species. Unlike any other species, we are aware that the world is a chaotic place. To deal with this chaos, we look for meaning. In the last chapter, we looked at higher meaning. With higher meaning, we search for order even if we cannot understand how it happens. With personal meaning, we are looking for order that can be created through our own activities or our role in contributing to order in the world, even if we play only some minor part in the total process. When we look for the meaning of life on an individual basis, we are looking for our purpose, identity, and security.

Change and the Need to Act

Life is about change. We experience change that takes place in the world around us, and we try to change things in order to make our life more comfortable and secure. To implement change, we must exert effort. Purpose provides us with both a goal to aim for and incentive to expend the effort. It is our ability as humans to anticipate future results of our actions that motivate us to engage in purposeful activities. People feel good when they achieve goals, but they can also feel good while working toward goals. On the other hand, when people must expend effort that doesn't appear to have a goal or they are not able to make progress toward a goal, they become frustrated. In other words, they feel bad.

Some day when we understand how consciousness works, we may be able to understand why purposeful activities make us feel good. Until that time, I am satisfied to believe that efforts toward a goal are similar to a basic drive to reproduce. We experience sex as a pleasurable activity, which is essential to preservation of the species. Purposeful activity is as essential to our survival as building nests are to the survival of birds.

Because we are self-aware, we can distinguish between ourselves and everything else. We give "everything else" a name or identity.

Likewise we must develop a meaning for ourselves in relation to everything else but, most importantly, in relation to our fellow humans. We fit into a society, and we must define our place in that society. Going beyond where we fit in society, we have a need to feel good about where we fit. Today, every individual is considered to have equal status and potential, at least at the beginning of life. During less enlightened times, all people did not have equal status, but they had a sense of where they fit in society nonetheless.

Finally, we live in a chaotic world. In reality, unpleasant things, over which we have no control, including death, can happen at any time. Fortunately, we have learned that by engaging in certain purposeful activities, we are able to reduce the impact of the randomness of nature. Growing crops is more predictable than hunting and gathering. Shelters protect us from the extremes of weather, and medicine can heal us in the case of sickness or accidents. In times of crisis, we have a need to *do something*.

Psychologists tell us that we also use meaning as a way of accepting unpleasant events. When an event appears senseless to us, we console ourselves with the idea that whatever happened must have happened for a reason, even if we do not know what the reason is. Experiments have shown that when people have only the illusion of control, the situation is more bearable. In one experiment, two groups of people were asked to sit in rooms with loud noises. One group was told that it could stop the noise if the noise became unbearable by pressing an escape button. The second group was not given an escape button. Although the first group's escape button did not work (but they weren't told this), the people in that group were less bothered by the noise than the people in the second group. Just the presence of the button gave the first group a feeling of control over the situation—they could escape if they needed to.

One final thought on change. It is human nature to desire stability. When things don't change, our daily lives are easier. Whenever possible, we develop habits to perform routine functions. To some extent, driving to work each day is part habit and part conscious activity. On the other hand, when life becomes too routine, life becomes boring. We need a balance between stability and change for a meaningful life.

In the previous chapter, the process of evolution was described as incremental adjustments to environmental change. Contrary to our intuition, complex engineering projects such as the human body can be accomplished by many small changes, given enough time. If we are going to look to what *is* as part of the process of establishing personal meaning, we should look at change as an opportunity to improve life. Many people look forward to lifelong learning. The optimist might say that in spite of all the benefits we have achieved through change in the last few years, we will most likely see even greater benefits in the future. In some cases, they may only remain opportunities, until we choose to take advantage of them.

As a result of social evolution, we now live in a more complex society than a few hundred years ago. Therefore, a need for a well-defined personal meaning is becoming more crucial.

The Trend Toward Personal Meaning

One of the few things that can be said with certainty is: over the last few hundred years, personal meaning has gained importance. Generally, the shift to personal meaning has been accompanied by a diminished importance in meaning that comes from religion. Before examining the trend toward personal meaning, it would be helpful to review the various types of meaning, all of which relate to the term meaning as purpose.

In the last chapter, we looked at the sources of higher meaning and what can be said about higher meaning at this time. The main purpose of looking at higher meaning was that one could assume that, if there is some higher purpose for life in the universe established by its creator, then we as a part of that creation could in turn as individuals assume that meaning applies to our lives, even if we played only a small role in the overall purpose of the universe.

We showed in the last chapter that it is not as intuitive as it would first appear that the universe must necessarily have a creator. At best, the most that the human species may be able to determine about the purpose of the universe is what *is*. While what is may not be adequate for some people, what is can be used as a reference in establishing personal meaning. Most people do not question what is, or what we have learned through science, when they seek to prolong life or make it more comfortable. When people become ill, they may say a prayer for their recovery, but they also take medications and undergo medical procedures, such as bypass surgery.

The first source of higher meaning was religion and myth. Higher meaning serves two purposes for religion and myth. First, it is an explanation of what is, even if the explanation is unclear or simply says that it is beyond human control or understanding. Second, it is an explanation of common purpose, which can be used for social control. Although the human species learned that it increased its chances of survival through social cooperation, religious organizations learned that social organization worked better if rules of behavior came with the authorization of a higher power.

Until the last few hundred years, self-interest was always viewed as being at odds with the common interest. Terms such as selfish and self-serving still have negative connotations today. Our original social organizations were basically extensions of the family. Kings or queens ruled the land, and they passed the reins of power down to their children. Regions of a country were ruled by brothers and cousins of the king or queen. Even in modern times, a family works for the common benefit of all members.

Eventually, social organizations grew too large to be governed effectively based on family structures. In medieval times, a king may have ruled a country, but day-to-day social organization was still basically at the level of a village. As trade and technology became more sophisticated, economic and political systems were developed that allowed individuals to receive an equitable compensation for their efforts or innovation.

The United States was founded based on the concept that the rights and self-interest of the individual were more important than the government or any other organization. The individual cedes to the government only as much authority as was required for it to govern. After the unprecedented economic success that we have enjoyed in the United States during the last 225 years, few people will argue that a system that allows individuals to pursue their self-interest within the bounds of the appropriate social organization eventually provides the most economic and political rewards.

As mentioned earlier, the trend toward individual self-interest has been a gradual but continual trend. In the early years of this country, individuals had political freedom, but religious organizations still held much influence over people—more than they do today. The United States may have been founded with the words "all men are created equal," but only recently have women, minorities, and the disabled been given equal rights in all areas of economic and political life.

One of the most recent manifestations of the trend toward individual self-interest is what has happened at major corporations over the last twenty-five years. At one time, there was an implied contract between a corporation and its managers and, to a lesser extent, with all employees. A company believed that its employees were a long-term resource. Companies recruited people and trained them, with the intention that they could make an investment in their employees and that employees would work long hours to improve the company's position in the marketplace. The underlying assumption in this contract was that basic market conditions would be stable or change slowly enough to allow a company to continue to provide its employees with long-term employment. However, today business conditions can change so rapidly that employees can no longer count on such an implied contract. Thus, self-interest is unavoidable on the part of managers and employees.

Thus, any discussion of personal meaning today is in a much different social context than it would have been only a few years ago. There was little sense of an individual thinking about personal purpose when a person's position in society was dictated by who their parents were or where they were born. People did have a purpose, but it was dictated to them by others. In some cases, people had a religious duty to lead a certain life if they expected to be a part of the community and eventually go on to a reward after death.

The trend toward the self has become so significant that people are now told that they should discover through introspection who they are and what is best for them. The self-help sections are the largest areas in most bookstores. Popular magazines devote many of their articles every month to topics such as self-actualization, developing higher self-esteem, and getting the most out of social and emotional relationships.

Relying completely upon oneself to establish personal meaning may not be entirely a good thing. A yearning for times when things were simpler is in part a desire for the comfort of tradition, when values and purpose were handed down from earlier generations. The pressure for self-direction and self-actualization is so strong that some psychologists say this pressure is a major cause of the increase in the number of people who suffer from depression today.

The Meaning of Life: Let's Not Talk About It

In many cases, the cause of depression is the fact that people are not comfortable with the goals and the meaning that they seem to be following in their life. In his book *Self Matters,* Phillip McGraw suggests that depression comes from people leading a life that is not in harmony with their authentic self. According to McGraw, there are a small number of defining moments, critical choices, and pivotal people that define the path that we follow in life and the satisfaction that we enjoy from following that path.

Some of these moments, choices, and people either provide us with positive input, or result in negative outcomes. Since at the time we get these inputs they may not appear to be life defining, we often are not even aware of the potential harm or good these inputs have on our efforts to be happy. And even if we notice these inputs, McGraw says, we still may be unaware of the impact on our behavior. McGraw's book provides readers with the tools to define the authentic self and then develop a plan that enables them to lead a life that is in harmony with their authentic self.

The popularity of McGraw's book and others like it indicates that people are more concerned than ever with a personal meaning of life. Still, we have trouble facing the issue, even talking about it. Surveys have shown that many people are uncomfortable talking about the meaning of life. Survey responses indicate that while many people may be aware that they are in some way missing out on the meaning of life, they are reluctant to discuss it. Even the media seldom tackles the topic. One of the few works in any media on the meaning of life is the Monty Python comic movie *The Meaning of Life*.

I believe part of the problem is that society doesn't recognize the importance of a personal meaning of life. The U.S. system of government allows for personal freedoms to pursue a personal meaning, but it is still influenced by institutions such as religion that are concerned with meaning and goals that contribute to the greater good of society. In many cases, these goals are not in conflict. Therefore, most of the lack of attention is by default rather than by any attempts at suppression.

One Life to Live

The boundaries on an individual life are confining. Few people live beyond one hundred years. In fact, the average person lives about eighty years. That eighty years is separated into sections that represent the cycle of life. The first twenty years, and in many cases today the first twenty-five years, are spent growing up and acquiring the skills to participate in society. Another twenty to twenty-five years are spent raising children. After a point, women pass their physical potential to have children.

Most people are born with average physical and mental capabilities. While everyone has the potential to excel, most people will be limited by the number of available opportunities. In many cases, opportunities will be further limited by timing—being in the right place at the right time with the right capabilities. Finally, in spite of the emphasis on the individual and the self in establishing meaning, people are still born into specific socioeconomic groups. Thus, even if we had a well-defined higher meaning of life, meaning for the individual could be quite different.

Meaning for an individual usually varies during life. The mid-life crisis and the empty nest syndrome are familiar names for situations in which meaning may be reevaluated or changed. Ironically, some people can work toward goals that are never achieved and that still provide meaning to their lives, while other people can attain goals and then find that their life has lost much of its meaning. As people's lives change, some of the things that can or do give their life meaning also change. Another irony is when we reach the end of our life, the meaning of our life is evaluated in terms of what we did for others—not for ourselves. Our obituaries mention the children and grandchildren that survived us, the charities we supported, the contributions we made to science or the arts.

Personal Meaning Is in Perspective

Fortunately or unfortunately depending on the circumstances, personal meaning is primarily a self-administered system, both in the process of establishing meaning and goals for one's life and in evaluating how well one has done at living a meaningful life. There are some people who would be identified by the vast majority of people as successful, but who personally feel that they have not achieved goals that they set for themselves or the goals achieved by parents or people whom they used as role models. On the other hand, there are people who may have had the potential to achieve significant success but are content to live simple lives with simple pleasures.

To some extent, it can be said that a meaningful life is a life lived in a good emotional state. People have a positive self-image when they feel their life is meaningful—that they are making a difference and that they are achieving their goals as a result of their personal qualities, which are grounded in a system of personal values.

People have multiple sources of meaning in their lives and are able to find meaning in one area when another area has fallen short of expectations. For example, many men will place an increased emphasis on family life if they have not achieved an expected level of success in work. In some cases, people will change careers, even taking lower compensation, to get more satisfaction or less stress. Just as more than 50 percent of college students are known to change majors at least once during their college years, most business people change careers at least once and sometimes twice in their work life. People are also known to make life changes as a result of events that would not be normally classified as a crisis or a failure to achieve goals. Generally, these events allow people to see their goals and purpose in a different perspective.

Many Sources of Meaning

Some people criticize our way of life in the United States as being overly materialistic, but U.S. society does provide individuals with many ways of establishing goals and achieving results. It is true that we consume a disproportionate share of the world's natural resources. However, our society also involves a system of science and technology that allows people to set goals related to the development of new products and services, and demonstrate self-worth in the achievement of these goals as they are used by people around the world. The high rate of technology development can be a double-edged sword.

The same technology that can provide a means of achievement for one individual also demonstrates that the new can quickly replace the old. This rate of change can instill a general feeling that old things soon become obsolete and, thus, promote a belief that traditional values are no longer appropriate.

The Internet is one example of how technology can have an impact on the meaning of life. The Internet has changed what it means to be a book retailer, stockbroker, and travel agent. Using the Internet, some small companies can compete on equal footing with large companies. More important, the Internet can change how we communicate with other people, and the people we communicate with. Although innovations like the Internet do not come along every day, there are other innovations that can impact the meaning of life.

We are just starting to scratch the surface with another technology, genetic engineering, which has the potential to have even greater impact on personal meaning. Although people mistakenly assume that cloning involves making an exact copy of ourselves, we may soon be able to change the genetic material that we pass along to future generations. We are already having an impact on our self-image through the use of cosmetic surgery. The ability to change our self-image is no longer confined to the clothes we wear or the things that we own.

We have many sources of meaning in our lives, which can lead to conflicts in establishing goals. Two of the greatest sources of personal meaning are work and family. We will look at both work and family in more detail in separate chapters.

We all know of the conflicts that women encountered when they first entered the workforce. Women tried to get meaning from their jobs in the same way and to the same degree that men do. Since women still bear more than an equal share of the childcare effort, they often found that they just didn't have the energy to do both jobs properly.

Women entered the workplace partly out of an effort to provide additional income, but also because they believed working provided them with an opportunity to achieve higher self-worth. Since women have always been responsible for the childcare and women have not enjoyed equal status with men, childcare did not provide the same status as participation in the paid workforce.

The efforts of women to participate in the workforce on an equal footing with men is a good illustration of how our feeling of self-worth can also be influenced by social values and standards. In other words, women feel their life is more meaningful, because they are able to participate in the workforce and they are able to do it in the same way that men do.

In recent years, and especially since September 11, 2001, people have been looking for more balance in the things that provide meaning in life. At one time, both men and women believed that hard work and long hours were necessary to achieve success. For men at least, working hard to provide the best for their family was a goal that provided self-worth and also made the effort feel meaningful.

A recent survey by Data Catalyst indicates that women are now seeking more of a balance between work and home life. In the survey, 70 percent rated companionship and a loving family as important, while only 20 percent said making a lot of money and becoming influential business leaders were important. In the 1970s and 1980s, when the last generation of women entered the workforce, success was defined by rising through the corporate ranks.

Women have paid a price for success. The study showed that only 67 percent of top women executives with MBAs were likely to be married, compared with 84 percent for men. In addition, 75 percent of the men have children, compared to 49 percent of the women. I suspect that the current generation of women saw the sacrifices their mothers had to make and the rewards they got, and realized that a balance was more important.

No Man Is an Island

The emphasis in this chapter has been on the self and how people determine meaning for themselves. However, we don't do this in a vacuum. We establish a sense of purpose in our lives and a sense of self-worth through our association or connection to other people. The two largest sources for meaning for most people, work and family, provide purpose as well as meaning because we are a part of a family, business, or nonprofit organization.

Our membership in the human species provides a great deal of meaning. Religion tells us that humans are different because they have a soul. Even scientists have some problem accepting evolution, because it says that we are just a part of the animal kingdom. Regardless of our religious or scientific belief, we assign a special value to human life. We already face moral dilemmas with regard to the death penalty and abortion, and genetic engineering will soon present the biggest dilemma we have ever faced.

We get various amounts of meaning from the country and city where we live. People are proud to be Americans, because the United States was established to follow certain moral principles. Many people choose to come to this country rather than continue to live where they were born. Most of the people living in the U.S. today are no more than three or four generations removed from immigrants.

The U.S. has traditions, although newer than countries in Europe or Asia, and accomplishments. Many people get a great deal of pleasure from attending or watching local sports teams. Today the greatest public celebrations take place when a local sports team wins a championship. Some people love the excitement of living in a city, while others enjoy the peace and quiet of the countryside.

People get meaning from whom their parents are. People can identify with the success of their parents even when their parents' accomplishments may appear humble, if they were accomplished against great odds or as a result of hard work. Unfortunately for some, a parent's success may be a burden, when a child feels that he or she must equal or exceed the accomplishments of a parent.

The Myths of Meaning

While I believe that the trend toward personal meaning leads to more meaningful lives and that establishing personal meaning should be based as much as possible on what we know is true, in other words, based on science, there are some myths with regard to meaning that we learn from society, that may also contribute to more meaningful lives.

The biggest myth is that life is mostly working toward goals rather than an experience in and of itself. To survive, society needs people to work toward goals that support daily life and the reproduction of our species. Based on our experience in the United States, society is more successful if individuals are motivated by a sense of well-being, a meaningful life, while they work toward those goals, but that sense of well-being is not essential. Other societies have used slaves or caste systems to get work done, while the leisure class enjoyed the experience of life. Christianity was formed during the hedonistic excesses of the Roman Empire. Thus, the goal of Christianity was to prepare people for the next life, which was their reward.

When Christian society gave way to secular society, material goals, and to a lesser degree a quest for knowledge, were substituted for the quest to save one's soul. If you ask a person what he or she expects in a life after death, the person will most likely say happiness. Thus, it is our Christian history that makes us believe that achieving material goals will bring us happiness.

A common myth about both the higher and a personal meaning of life is that there must be some sense of permanence—that the rules must apply all the time in all cases. It is easy to make such an assumption, since many of the laws of nature such as the laws of physics have remained unchanged for billions of years. However, as is shown in the chapter on uncertainty, there is much that is subject to randomness. This randomness leads to change.

The one common factor in life over time is change. Nature changes and people change. Our bodies change and our relationships change. Ironically, the concepts of higher meaning, such as justice or creation, which one would expect to be more stable, are in fact the concepts that change the most over time.

In general, we expect the current state of affairs will continue indefinitely into the future. When we are young, we really have no concept of death. Every couple that gets married expects their marriage will last until death do us part, although more than half of all marriages eventually end in divorce. Although life involves change, life also tends toward stability. Our bodies maintain a current temperature, and we have a constant blood pressure. The process of evolution itself is a process of change to keep life in balance with the current environment.

It is a myth that the universe exists just for humans. Many religions teach that the earth and all life on it besides humans were created for humans. Ironically, while science can make a case for the fact that the human species is nothing more than a step in the evolutionary process, we do find ourselves in the unique position of having more control over life on earth than any other species. Most important, due to our large brain size, we have the capacity to be self-aware. Science believes that due to the size and age of the universe, the potential for self-aware life elsewhere in the universe is likely.

An essential component of conscious experience is the concept of our being separate from our environment. Psychologists tell us the capacity to self-identity does not appear at birth, but only after a child is about two years old. Thus, in the sense of our conscious experience, the universe was created for every one of us. Each conscious being's experience is a unique experience of life or reality.

It is a myth that there must be an answer to all questions. Again ironically, science, which works with the principles of cause and effect, leads us to believe that there must be an answer to all questions. The laws of nature lead us to believe that all things must fit into some master plan, even if we cannot ever know what the plan is. However, science has discovered events in nature, that it has yet to define even an approach to finding an answer.

If we assume that we are individuals with self-determination, we still reserve the right to accept or reject any of the above concepts. However, I believe that in establishing personal meaning one must at least leave oneself open to the possibilities of questions that have never even been considered.

Chapter Seven

Work, Occupation, Career: We Are What We Do

"I keep reading stories about CEOs of large companies who make hundreds of millions of dollars in stock options. There is some debate as to whether this is appropriate. One argument is that these CEOs are visionaries, uniquely qualified to create spectacular shareholder value. Another possibility is that CEOs are just showing up and shuffling things around until something lucky happens. I'm leaning toward the 'showing up and shuffling' theory.

"I'm not saying CEOs are dumb. Put yourself in their shoes. When you're a CEO, the only information you have is what your subordinates give you. And they're all unscrupulous sycophants. The last thing you'd ever hear is the truth. So there you are, a powerful CEO astride some mammoth enterprise, armed with no useful information whatsoever. You know you have to do something, but there's no way to know what. Your only rational strategy is to do random things until something lucky happens, then take credit."

— Scott Adams, Creator of Dilbert

Work Provides a Sense of Self-Worth

Work both gives meaning to our day-to-day activities and provides a sense of self-worth. Also since the income we earn through work enables up to sustain our life and the lives of the people in our family, it gives us a sense of accomplishment or making a difference. The meaning any one person gets from work varies. At one time or another, we have all have been asked, "Do you work to live, or do you live to work?" For workaholics, work is all the meaning they need in life.

The concepts in this chapter are closely related to those in chapter eight, which deals with sex, family, and relationships. The sense of meaning and satisfaction that individuals get from a family relationship is closely related to how well they provide for their family's financial needs. At one time, the primary factor in selecting a mate was the potential partner's religious beliefs. Today, people establish relationships based on what their potential partner does for a living, and how well they may be able to provide for family needs.

This chapter deals with work or, as the Internal Revenue Service calls it, occupation. All the work that we do, in the traditional sense of the word work, does not result in income. Many women list housewife or homemaker as their occupation on their tax forms. By far, the most important factor in any person's concept of personal meaning is work. Some people would say that work is the most important aspect to personal meaning, because it is the thing we spend the most time doing. Rather, I believe that it is the most important aspect to meaning, because it defines what our function is in society.

When two people meet for the first time in a social setting, it is likely that within the first few minutes the conversation will turn to what one or both of the people do for a living. In every obituary, there will be a mention of what job the person held, even if he or she was retired for more than twenty years before he or she passed away.

In today's economy, money provides a common medium of exchange. So, it can be used to put value on things. It has often been said that when talking about the million-dollar salaries of sports stars or corporate executives, money is more of a way of keeping score in a competitive world than providing pay for services rendered. Actually, it is a way of keeping score for everyone, at least to some degree.

While people will spend their income on things that will demonstrate their general level of wealth, all people, poor and wealthy alike, are reluctant to discuss exactly how much money they make. Employers often discourage employees from disclosing their salary. This is an easy policy to follow since most people don't want to know if they are making less than their fellow workers. If your employer thinks your fellow worker is worth more than you, that's a blow to your ego. On the other hand, if he thinks you both have the same potential and he is still paying you less, it's also a blow, since you weren't as good a negotiator as your higher paid coworker.

In today's economy, salary is more than just compensation for a person's labor. Salary is compensation for a person's services, and a person can now get a sense of personal worth based on the services he or she can provide. Today young people spend much more time preparing to enter the workforce. While most people go to college to get a better paying job, people also realize that the jobs college graduates get after school have more prestige—thus adding to a person's sense of self-worth. Life is more meaningful for people with so-called professional jobs, especially in their early years. Trades people can earn more money than recent college graduates, but these jobs hold less prestige, even when the tradesperson is operating his or her own business.

Ironically, key corporate executives, professional athletes, and entertainers are paid millions of dollars a year for doing things that make their lives meaningful. Meanwhile, the lowest paying jobs today provide the least amount of meaning, because we have been so successful in automating most routine jobs, which are often also the most unpleasant.

Many of the lowest paying jobs are in retail, food services, and domestic services, which offer limited opportunity for advancement. The limited opportunity for advancement also limits the personal meaning one can gain from these jobs.

The Changing Nature of Work

Webster's Unabridged Dictionary lists twelve different definitions of the word work and another thirty-nine definitions of applications of the word work or works. In spite of our understanding of the work ethic, some people might be surprised to learn that the idea of working for pay was once considered a bad thing. At one time, Christianity believed that hard manual work was good for the soul, because it allegedly helped people suppress their sinful inclinations.

The Work Ethic

At the beginning of the industrial revolution, society needed a way to encourage people to work at jobs that provided little or no intrinsic satisfaction. Thus, society developed the concept of the work ethic. The work ethic taught that hard work and perseverance were the way to get ahead. Hard work also built character. At least it built qualities that were good for employers. However, before our modern economic system, most people did not work for wages. People worked on a farm or in a family trade. Working was a way of supporting and participating in the community. Thus, working for money was considered a sin of selfishness or a form of greed.

Work has not always been a source of self-esteem. At one time, work, which was defined as manual labor, was only performed by the lower classes. One of the ways of defining an English gentleman was a person who did not have to work. An English gentleman simply managed his affairs, which did not take much time in those days. Although it is easy to assume that the work ethic always existed, one could say that the work ethic is an American value. When the first settlers came to America, life was hard and people lived on the edge of survival. Everyone had to work in order to survive. Thus, a leisure class, people who by definition did not work, never developed in the United States. However, some people did have to work harder than others.

We look at managers in business today and it is easy to assume that management has always been hard work. Managers today are usually the people who put in the longest hours; they are the ones caught in a trade-off between work and family. To hold a job in today's business environment, one must put in so many hours at work that people do not have enough time to spend with their spouse and children. However, before the days of rapid or instant communications, which was only about one hundred years ago, there were a limited number things that a business manager could manage. In many businesses, all useful management activity could be done before lunch.

Careers vs. Jobs

The hardworking manager of today is a result of the transition from simply working at a job to having a career. Today people start working on a career even before they get to the workplace. A college degree is now an entry requirement for many jobs, and a Ph.D. is required to do any meaningful work in any of the sciences. Doctors and lawyers also require several years of additional education beyond an undergraduate degree, before they are allowed to practice their profession.

Once people get into the workplace, many entry-level employees in fields such as law and accounting are expected to work much longer hours than partners as a way of demonstrating a desire to be worthy of becoming a partner. The people who do not make the partner level by a certain age usually leave the firm to find another path to success. The system has often been called the move up or out system. The transition of work from a job to a career has created additional meaning to work. A career now provides a goal, which can provide meaning to a person's efforts in school and on the job.

In the United States, there never has been an elite class that realized its status simply by birth. Working in business has been the primary way to establish one's status in the United States. It is not surprising that the homes of the rich in the early part of the twentieth century were modeled to reflect the sense of status implied by the castles and manor houses of European nobility.

In recent years, American business has seen a shift in the idea of a career. At one time, people believed in a career concept that encouraged hard work on behalf of the company. In the 1950s the organization man in the U.S. and more recently the company man in Japan devoted his efforts to working his way up the ranks of a single company. This system included the legends of the person, exclusively a man, who started in the mailroom and worked his way up to the president's office.

Due to globalization of markets and the rapid change in technology, which has enabled companies with improved products to replace longtime market leaders, companies no longer have the ability to provide lifetime employment. Thus, working for a specific company and identifying with a company's products has less ability to provide meaning to people today.

Today, we have a new work ethic, which involves bringing values to the job rather than getting value from hard work. Managers are expected to show loyalty and work as team players. Jack Welch, the former CEO of General Electric, is considered one of the best managers of the last twenty-five years. When talking about managers at GE, he has said that there are two kinds of successful managers: the manager that produces results and is a team player and the manager that produces results but does so using his own methods, which are not always acceptable. Welch intimated that eventually the second type of manager is asked to leave, but it is a difficult management decision since the manager produces results, which is the primary driver in business.

This identification of work with meaning is one reason why most of the efforts in the women's rights movement have focused around the workplace. Traditionally, women performing similar jobs as men were paid less than men. Women also have had fewer opportunities to move up through the corporate ranks. All measures to eliminate these inequities ultimately will provide a better sense of self-worth for women.

The Work Ethic and the Working Poor

If the work ethic is dead, as some sociologists insist, because workers have realized that hard work in unpleasant working conditions does not lead to success and a feeling of higher self-worth, then why does society continue to base its work programs on the work ethic?

Most of the welfare-to-work programs were justified on the basis that if society could get people off welfare and into the workforce, they would eventually be at a higher income level and, more important, have a higher sense of self-worth and purpose. According to the work ethic, a working person is better than an idle person even if he or she has no more income. However, if the primary purpose of work is to provide for one's livelihood, it is difficult for people to leave social welfare programs if the net effective income of the working poor is less than the welfare benefit. According to the cynics, the welfare-to-work programs were just a ploy to reduce public spending.

There is no question that welfare programs that just hand out money without any plan to break the cycle of one generation after another receiving benefits must be changed. However, moving people into the bottom end of the workforce may not be the right solution, until some changes are made to this sector of the labor market. In addition, there are many other people who have never received public assistance, who must try to make a living wage through jobs in this sector.

Since there are such a large number of people who must participate in the lowest end of the job market, I would like to look at jobs in this sector from the perspective of the meaning of life. It is often said that when people are engaged in survival, they have no time to think about the meaning of life. I do not believe this is true.

In her book *Nickel and Dimed: On (Not) Getting by in America,* Barbara Ehrenreich detailed her experiences as a member of the working poor. During a period of several months, she worked as a waitress, for a home-cleaning service, and in retail at what would be considered the lowest end of the labor market. Ehrenreich wanted to determine if a person could earn enough money at the lowest end of the labor market to provide for the basic necessities of life: food and housing. After all, providing these necessities is the primary objective of work.

Ehrenreich, who worked in several different geographic markets, reached the general conclusion that, depending on the cost of housing in the specific marketplace, the working poor can earn just a little more or just a little less than is required for basic survival. Basic survival was defined as taking any housing that was available that would provide basic shelter and security, and either preparing simple meals at home or purchasing fast food. Basic survival did not include any form of healthcare or automobile expense. Ehrenreich discovered that if the working poor incurred any other unavoidable expense, or became ill and could not work, they quickly fell behind financially. She pointed out that most people who work in this sector hold two jobs, if they have the physical strength and stamina to do it, or they team up with another individual to help share expenses.

Ehrenreich worked undercover in the sense that she did not let people know, including her employers, that she actually held a Ph.D. and that she was working in these jobs to gather material for a book. Thus, her coworkers shared their feelings, frustrations, and attitudes with her as if she was someone who was in the same predicament.

Based on her account, it appears that many of her coworkers still believe in the work ethic in spite of the fact that many employers indicate through their policy and procedures, that they don't really believe in it themselves.

Ehrenreich says that as a waitress she was told that her boss could search her purse at any time. Almost all jobs required mandatory drug tests. In other words, a person could not sm . start until the drug test report was negative. Many jobs required people to take personality tests, which in addition to questions such as "would you turn in someone you saw stealing?" also asked questions about their moods and feelings. Although employers treat people in low-level jobs with suspicion, Ehrenreich claims that she did not encounter any thieves or drug addicts among the people she worked with.

Most importantly, employers did not provide a free and open market with regard to wages and opportunities. Many employers forbid employees, under the penalty of dismissal, to disclose their pay to fellow employees. When applying for jobs, applicants were often given misleading information about starting wages and the ability to get raises. Naturally, employers will deny that these practices exist, but they also know, but are not willing to admit, that if they and not their competitors allowed wages to follow the laws of economics, they would eventually go out of business. The bottom line is that it is difficult for people to improve their position through hard work, and they have little opportunity to develop any sense of self-worth and purpose in life from these jobs.

Many years ago, when I was a student, I worked in low-end jobs as well. These summer jobs were one of my biggest incentives, when I was back at school, to study hard. However, everyone does not have the opportunity to move high enough in the job ranks to avoid the practices by employers that take any potential for personal meaning out of a job. I believe that we should limit our expectations with regard to the meaning that we can get from working. However, as a general rule, society will benefit from an effort to put the most meaning possible in every job.

According to Ehrenreich, the management style of first-line supervisors and company policy were to discourage any form of initiative from workers. In spite of their managers and company policy, many workers were able to find meaning in their work by taking the initiative at times. In general, she believed that many workers succeeded in spite of, rather than because of, management direction.

Work as Creation

The word work can be used as a noun as well as verb. *Webster's* also defines work as "a product of exertion, labor, or activity: *musical works.*" Work also provides personal meaning when a person can identify with something that is the result of his or her efforts, skills, or intellect. Many people today engage in cooking, gardening, or woodworking simply because their jobs don't involve a tangible product. These people work on projects that involve hundreds or thousands of people all engaged in a common effort such as developing a new electronics product or a communications network. While people can identify to some degree with such products, it does not provide the same meaning as building a house or making a cabinet.

A few years ago when Japan was enjoying so much success selling its products in the United States, many studies were done to understand why Japanese manufacturers were being so successful. One reason was that the quality of Japanese products was higher or at least the consumer perceived them to be higher. The high quality of Japanese products was attributed to the Japanese approach to work. Although U.S. businessmen spent long hours on the job, it was well known that Japanese businessmen spent even longer hours on the job.

To quantify the difference in hours worked, a sample of U.S. and Japanese business executives were asked to keep logs on how they spent their time. I no longer have the details of the study, but I do remember the important findings. The study showed that Japanese businessmen did in fact spend more hours on the job—about ten hours more per week. However, the data in the logs also revealed an unexpected fact about how time was spent by both groups.

Both groups performed some work not related to their jobs, such as projects working around the house, gardening, or woodworking, that would be classified as work for pay if it was performed by someone else. The U.S. businessmen spent a significantly larger amount of time on these activities. The net effect was that U.S. businessmen spent just about as much *total* time working as Japanese businessmen.

Motivations to Work

The findings of this study points out the need to take a close look at the nature of work and the nature of the workplace in the U.S. to fully understand the significance of work to the meaning of life. *Webster's* also defines work as "employment, as in some form of industry, esp. as a means of earning one's livelihood." Among the adjectives that *Webster's* lists are drudgery and toil. Today, a large number of people would have to say that this definition and its adjectives are contradictory when thinking about the work they do. In addition, while most people must continue to work as a means of earning their livelihood, many would continue to work even if they did not need to work to support themselves or their families. Many lottery winners, if they had a job before they won the lottery, continue to work, even if the lottery money is sufficient to provide for their livelihood. This indicates that money is only one motivation for people to work.

There are many activities that one person does for pay that another person does for recreation. Besides the obvious examples in sports and entertainment, many people participate in community services as volunteers, work on home improvement projects, or even build boats and airplanes. Thus, most jobs have some aspect that people can point to that provides meaning to their life, either by providing purpose or adding to self-worth.

Trends in Meaning from Work

As mechanization improved the productivity of the average farmer, many farmers left the farms to work in factories, either for good or during times of the year when they were not required on the farm. Most early factory jobs involved hard, monotonous work in unsafe and unpleasant conditions. It was during this period that toil and drudgery were appropriate adjectives for work. However, even under these circumstances, workers made efforts to take control of their situation. Having some sense of control over the process made the work more meaningful even if it did not contribute to a person's sense of self-worth.

People working on assembly lines would make efforts to control the pace of work, even if it simply meant working faster for a period of time so that the parts made during a busy time could be used to feed the assembly line during rest periods. Today, some auto manufacturers allow workers to stop the assembly line when a problem develops that would affect the quality of the cars on the line. Fortunately, many manufacturing processes have been automated, and many manufacturing jobs involve managing the flow of work rather than actually working at repetitive tasks.

As jobs were eliminated in manufacturing, these people moved into service functions. Once again these jobs, which could be in the front office of manufacturing companies or in service companies such as banks and insurance companies, involved a lot of paper shuffling. Many of these jobs, such as key punch operators, allowed less control than assembly line jobs. These jobs could be even worse than assembly line jobs because they provided no tangible contact with the product or service being provided by the company, which did not allow a person to identify with the product or service. In the service sector, many middle level mangers also spent much time reporting and consolidating information on production to senior management.

With the introduction of information technology, many of these jobs were automated and the people were laid off or moved to other positions. Again fortunately, for the people that remained, their jobs become more meaningful. People usually had a larger scope of responsibility, and they spent more time analyzing and planning future activity, which can provide a higher sense of purpose and, in many cases, higher compensation.

Finding Meaning in Self-Employment

One of the benefits of information technology for business is more information sooner. Companies now have information about how product is flowing through the manufacturing process and distribution channels almost on a real-time basis. In the past, companies would have to put production schedules in place and ship product to customers according to a plan with only periodic updates about how actual results were going compared to the plan. Without this information, companies had little capability to adjust production to accommodate variances from the plan. However, even if the information was available, it would only be useful, if a company had the ability to adjust the use of resources. Thus, with the availability of timely information, companies are seeking to gain flexibility in their operations, especially in the areas that would be considered support functions.

Business analysts predict that the company of the future will have a small core group of people, who will manage vital functions. The remaining functions will be outsourced to consultants and support companies. The net result of this trend is that the labor force will be composed of more self-employed people and people who work for small companies, where they will be owner operators.

Many people who work dream about being their own boss some day. People express it in terms of being their own boss rather than being self-employed because they yearn for the control that they will have over their future, even if everything depends on their actions. Having worked for major corporations and also having been self-employed for several years, I have experience with both the good and bad sides of self-employment. Based on my personal experience and the experience of many other self-employed people, life is more meaningful as a self-employed person. Thus, one can say that if the trend is toward self-employment, it is also toward getting more meaning from work.

Limits to Meaning from Work

Whenever people think about work, they almost invariably think about the businesses that operate in the for-profit sector of the economy. In reality, more people work in the not-for-profit sector than in the for-profit sector. Nonprofits include government, private education, some healthcare, churches, and non-church-related charitable organizations. The not-for-profit sector provides an opportunity for people to find meaning from work, especially if they choose this work specifically because they believed that this work could provide meaning.

There is a small number of people who work in the nonprofit sector that are engaged in a special kind of work. This work is often referred to as a calling. People in religious organizations are the best example, but some people in medicine, education, and non-religious charities will also say that they are following a calling. While most people will say that the calling is from a higher authority, some people say they are following some inner feeling that they should do some particular work. In some cases, these people could even be reformed criminals. In any case, in terms of meaning, following a calling is like a self-fulfilling prophesy. Following a calling can be a meaningful life. However, if a person becomes disillusioned with a calling, he or she usually makes radical changes in his or her life. There is no half-way in terms of meaning in the case of a calling.

Since organizations in the not-for-profit sector are not set up to earn a profit, one might assume that work in the not-for-profit sector is more meaningful, but the daily activities of most people in this sector are not that much different from the activities of people in the for-profit sector. Thus, one should not assume that work in nonprofits is any more or less meaningful than work in the for-profit sector.

One of the major limiting factors of achieving lasting meaning from work is the fact that most organizations are dedicated to change. Companies want to bring out new and improved services. Every new political administration wants to eliminate or change programs introduced by the previous administration. Thus, any mark a person may make on an organization is soon changed by the efforts of new people to make things better. Computer hardware and software provide some of the best examples. Companies, such as Lotus Development or Compaq, which were the pioneers in the industry as little as twenty years ago, have been replaced or absorbed by other companies. AT&T built the best telephone system in the world, but the company's name is in danger of disappearing as parts of the company are sold to competitors. It is part of the evolutionary process to improve things, but people must be content with being a step in the process.

It is also difficult to find meaning in fluid work situations. Organizational changes can lead to layoffs. In many cases, people are laid off en mass, and these lay-offs have no connection to a person's capabilities or anything else that the person could do to influence the decision. With lay-offs, people experience both a reduction in earning power and a feeling of having no control over their destiny.

The Self-Reinforcing Loop of Meaning

The topic of fulfillment is applicable to several chapters of this book, but I believe the heart of the problem stems from the workplace. The problem is this: how do we find personal fulfillment in a business environment geared to selling more and more goods and services, which could in turn lead to additional problems in energy use, the environment, and geopolitics?

With few exceptions, every business manager has two overall objectives in common with every other businessperson regardless of the product or service sold by his or her company. The first is to increases sales by a specified amount every year. A desirable increase might be at least 10 percent. The overall economy doesn't grow at this rate, but every manager knows that for a company to be successful and for the manager to be considered a good manager, a company should aim to be a market leader, which means growing faster than the overall market.

The second objective is to maximize profits. In most industries, a company can maximize profits by being the number one provider in the market. The number one provider is usually the one that sells the most units and is able to keep its costs low, due to economies of scale. Thus, it rakes in the most profits. But to do this, everyone focuses on selling as much as possible.

In the early years of the industrial revolution, the focus was on production and reducing cost. As people moved off the farm and into the city, they had basic needs for food, clothing, and shelter. With the invention of the automobile in the beginning the twentieth century, the automobile was soon seen as a necessity. As a result of mass production, consumer goods were relatively inexpensive and consumers needed little incentive to buy what would be considered the basics of life.

While there are still some poor people in the country today, after the Second World War one could say that the demand for basics was satisfied. As new consumer products were developed, such as color television, personal computers, and dishwashers, these products were added to the market basket. Therefore, since the Second World War, there has been an increased emphasis on product marketing. Marketing executives started to study the purchasing habits of consumers and their motivation for buying products. Most products are now sold more on their perceived benefits than their actual benefits. The marketing wars between the cola companies is a good example. The cost of packaging and marketing of consumer products in most cases far outweigh the cost of the goods in the package. Style is the most important factor in the sale of most clothing as well as automobiles.

People have been conditioned through advertising to believe they are defined by the products they buy and use. Although there is emphasis on quality, more and bigger are still major sources of ego gratification. At the beginning of the twenty-first century, consumer marketing has had a significant impact on what people believe is the personal meaning of life. We get a sense of self-worth from the products that we are able to buy, and our purpose in life is to acquire as many goods as possible, which will provide us with some anticipated happiness.

In spite our image of ourselves as hard workers, people today spend fewer hours on the job than a hundred years ago. Even executives who might put in more time during the weeks that they are working get more days of vacation. Today, people are told to stop and smell the roses. Part of this trend toward more leisure time is to allow people more time to spend the money they make. This increases consumption. In the most recent recession, economists noted that consumers continued to spend, and that their spending actually shortened the recession.

I present this case not as criticism of materialism, but simply to show that the U.S. economy is involved in a self-reinforcing loop of people trying to find meaning. Executives find meaning in their life by successfully competing in an economy that is fueled by selling products to consumers on the basis that their products will make their life more meaningful. If people are concerned about the overuse of natural resources by the U.S. economy, it will be necessary to find something other than compensation for increased sales as the primary indicator of executive worth. Since I am not suggesting that we move away from a system of free enterprise, profits will always be a motivator.

The Trend Toward a Simple Life

In spite of all the marketing hype, it appears that consumers are starting to feel overwhelmed by the gadgets in their lives as well as all the demands on their time. Ironically, business is using this desire for a simpler life to create a new business that will help people achieve a simpler life but not necessarily a more austere life. There are now magazines, Web sites, and books dedicated to helping people engage in selective indulgence. However, none of these resources are really discouraging consumption. The magazines and Web sites depend at least in part on advertising revenue. Thus, they depend to some extent on selling products for their success.

This recent trend toward simplicity is nothing new. Henry David Thoreau addressed this issue more than 150 years ago in his book *Walden*. Thoreau spent two years living a subsistence existence on Walden Pond from 1847 to 1849. His book, which was based on this experience, was published in 1854. Most people today would consider the life of the average citizen in the middle of the 1800s, not the life that Thoreau lived in the woods, as simple. In any case, even in those times, Thoreau was able to consider the additional demands that possessions place on the individual. However, while many people point to the book as an excellent literary work, it does not appear that his work has had much impact on other people's thinking with regard to the meaning of life in the last 140 plus years.

The Coming New Leisure Class

Regardless of changing attitudes toward the value of work of any kind, the changing demographics of the U.S. population will result in a larger proportion of the population being out of the labor force. Perhaps this new larger leisure population will eventually change our attitudes toward the value of work as a component of personal meaning. Some historians blame the decline of the Roman Empire to a large leisure class. One must keep in mind that all the work performed in Roman times was done by other humans or animals. Today much physical work is done by machines.

Most of the manual work that must be done by people is done by people who are paid low wages. The most important factor preventing the automation of many of the remaining manual labor jobs is the fact that there are sufficient people who will still work for such low wages that it is not yet economical to automate these tasks. However, I believe that within the next two or three generations, the number of man years required to provide a high standard of living to our population will go down. At the very least a larger proportion of the population will be engaged in activities related to the arts, science, and recreational activities.

Just as the upper classes in society in the past have helped the less fortunate people within their own country, more affluent nations will help the developing world to improve its standard of living. In many cases, the progress will not come from charity, but through efforts to help others help themselves.

Chapter Eight

Sex, Family, and Relationships

"I did not have sex with that woman, Monica Lewinsky."
— William Clinton

"The heart has reasons that reason cannot comprehend."
— Blaise Pascal

That Pesky Sex Drive

What would a book be without a little sex? Kidding aside, sex directly and indirectly has more impact on finding meaning in life than anything else. Sex will also have more impact on our journey through life than any other factor. Ask Bill Clinton, former president of the United States and leader of the free world, whose political legacy will be forever tainted because of the impact of his sex drive. Bill Clinton's behavior would not have had near the personal consequences as it did if our society did not place such a significant emphasis on sexual behavior.

Some people complain about explicit sex in the movies today and in the media in general. Some people may have the impression that there is more emphasis on sex today than there has been in the past. It is probably true that we are more open about discussing sex and even making efforts to enjoy sex more today than in the recent past, but sex manuals such as the *Kama Sutra* have been around for centuries. Our basic sex drive has always been strong, because it is essential to the continuation of the species.

We all should be thankful that our sex drive is so strong or else we would not be here today. Human reproduction is difficult and giving birth to children and raising them is a risky process that involves a huge investment. Part of the price that the human species pays for its dominance of the environment is the period of time we must spend growing to maturity. The human species has by far the longest period when young are dependent on their parents. As a result of evolution, it takes human young twelve to fourteen years to reach physical maturity, or the point when reproduction is physically possible. As a result of social evolution, it takes twice that long before humans reach the point where they are economically independent and prepared to marry and raise children.

Some scientists have coined the term "selfish genes" to emphasize the importance of reproduction and the continuation of the evolutionary process. In other words, the only purpose of life for any species is survival long enough to reproduce. Everything in life is focused on this end.

In spite of its importance, sex does not receive sufficient and unbiased consideration when considering the meaning of life. In western society and especially in the United States, we are still dealing with the excesses of the Roman Empire, which inspired our negative attitudes toward sex. The most extreme attitude toward sex takes the position that sex is only appropriate when it leads to reproduction. A large majority of the people in the United States believe that only sex within the bounds of marriage is appropriate. All other sexual activity is either sinful, illegal, or both.

I do not intent to take any position on which sexual activities are sinful, illegal, or inappropriate, but I would like to emphasize its importance to anyone who is attempting to develop perspectives on the meaning of life. Our sexual drives are as basic as our need to breathe, eat, and drink. Breathing, eating, and drinking maintain the individual. Sexual reproduction maintains our species. Sexual reproduction involves the oldest and most basic social unit in any society, the family. Two of the Ten Commandments relate to family. A large part of civil law relating to property specifically considers the family as well as the individual.

In spite of the fact that we repress our discussion of sex and hold different attitudes about its importance, its influence is continually present in our lives. Everyone knows that sex sells. Even when sex is not used subtly or explicitly to sell products, marketing campaigns still sell products by appealing to our desire to attract or be appealing to the opposite sex. Marriages and the formation of family units drive the market for housing, furniture, and appliances.

Religious leaders and social scientists may express concern over the divorce rate and the decline of the basic family unit, but most divorced people remarry and the birth rate is still about equal to the replacement rate for a steady population. We may have a problem with our basic social units in society today, and we may not be raising children in ideal family settings, but we are still reproducing.

Sociologists tell us that as a specific segment of society becomes more affluent, the birth rate in that segment declines. In poor economies, children are expected to contribute to the household income, and with extended families living in the same household, children often care for their parents when they are no longer able to provide for themselves. In advanced societies, the family no longer operates as an economic unit in the sense that it no longer produces goods, such as on a farm. Even at the beginning of the industrial revolution, but before child labor laws, children could work in a factory and contribute their earnings to the household.

In advanced economies, additional children in a household represents an additional financial burden, because they require more education and spend a longer period of time living off the income of their parents, before they can be self-sufficient. With the decline in extended family units, people have more income to save or invest for when they no longer work. In other words, people in advanced societies have new options for investments that will provide for their old age.

Sometimes changes in economic conditions can be confused with changes in social values. Some people might say that the decline in the extended family has a negative impact on society, but I wonder if the ability of older people to be more active and self-sufficient after they stop working is a positive factor that more than offsets the decline in extended families. Some people may be surprised to learn that at one time as high as 65 percent of adult women remained unmarried. These women were responsible either for the support of their parents or they could not find men who were willing or able to be responsible for their support.

While we may be able to downplay the role that sex plays in our lives for puritanical reasons, sex still has an impact on our daily lives, regardless of how much attention we pay to it. We can repress or deny our sexual impulses in an effort to comply with religious teachings, but in the long term, the behavior will be a serious impediment to the process of establishing a balanced set of goals, which when achieved provide satisfaction and meaning in life.

A few people will become senior corporate executives, or hold political office, or achieve recognition in sports, science, or the arts. But most people will live ordinary lives where family and home will be the most satisfying thing in their lives. Even people who achieve what may be considered extraordinary success in their vocation still place a significant value on family.

In the previous chapter, we talked about the conflicts that can arise between the demands of a career and spending quality time with the family. The issue is primarily a matter of balance. We only have so much time in our lives, and people must be able to balance options. As golf and football widows can attest, work is not the only thing that throws off the balance of the family; recreational activities, to a lesser degree, can cut into the quality time of the family as well.

The Role of Family in Establishing Meaning

Our first source of information about the meaning of life comes from our parents. We get this information both from what they say and how they act as our first role models. By the time we are ten years old, we have been given basic information on such issues as is there a God and how we get by in the world. We also have some idea of our socioeconomic status. We know if we are rich, middle class, or poor. If we are born into a lower socioeconomic status, where providing for basic needs is a struggle, we are going to have a different concept of the meaning life than a child that is born into a family that easily provides for his or her every physical need. In general, our parents' values and the value of community in which we live are going to be our values. How effective these values seem to be—do they enable our parents to live happy lives?—will determine how we set our own values.

Psychologists can explain in detail how we go through this process as we grow up, but the key point here is that we really don't start with a clean slate when we begin as adults to attempt to find some meaning in our lives. As children, our immediate focus with respect to a purpose in life is to grow into adulthood and acquire the skills to live as productive and self-sufficient adults. While we are growing up, we do not have many alternatives. We are in a dependent position and must accept the values of our family in order to survive—at least until we are able to provide for our own needs.

As adults, family has a huge impact on our process of establishing meaning, because we expend the majority of our adult time and resources providing for the needs of our family.

It is not surprising that children practice the same religion as their parents. In fact, it is unusual if they reject the faith of their parents and practice another religion. Practically all religions encourage, if not require, that their members marry people who are of the same faith. If marriage to a nonmember is permitted, the religion almost always requires that children be taught to practice the religion of the member.

Children follow their parents' examples in other areas as well. Many children select a career that is the same or related to their parents' careers, even when a family business is not involved.

Accepting the values of the family or an ethnic group is imperative to maintaining an affiliation to the family or group. Adolescent children often temporarily reject some of the values of their parents, as a way of establishing their own identity. However, rejecting the basic values of the family or an ethnic group may require discontinued contact with the family or group. This is an extreme price to pay to follow alternative values and meaning.

Since family and ethnic affiliation can in itself be one of the most important factors in providing meaning to life, it may be difficult for some people to even begin a process of exploring values and concepts beyond those accepted by the family or ethnic group to find meaning in life.

Dysfunction

There is no question that much that is good about us is passed along to us by our parents. All parents want the best for their children. Therefore, no one would intentionally pass along attitudes or traits that would make our lives less meaningful. Still, it does happen.

Psychologists tell us that much of our dysfunction is acquired from our parents. If our parents lived in a dysfunctional environment when they were children, it is repeated when they raise their children, and their children repeat the process with their children. Dysfunction can be passed on through generations because many families are not aware of the cause of their dysfunction. Thus, we must recognize that when thinking about the meaning of life, we may unconsciously be including dysfunctional ideas. Our ideas about the meaning of life cannot be completely free of bad information. By the same token, we inherit social biases. Since we have benefited from social evolution, it most likely is not a bad thing to include social biases in any meaning of life we develop, but we should make every effort to be aware that these social biases are present.

Other Communities of Thought

That major point of the chapter on religion, philosophy, and science was that, over time, we have taken different approaches to establishing meaning. One approach has replaced another. For example, science provided a different explanation, than the one provided by the Bible, of the position of the earth in the universe. However, within the scope of science, scientists are also subject to social pressure with regard to thinking on concepts that relate to the meaning of life.

Practically all science today is practiced by people who are affiliated with professional societies and who must be certified as qualified by earning either a master's or doctoral degree. Only people with these degrees are given the funds and resources to conduct scientific investigation. Both the process of getting the degree and the publication of new research are subject to peer review. In other words, new information and theories are subject to the review of people who have already been working in a branch of science for several years.

The process of peer review is an important part of scientific investigation. It assures that people do good science that follows certain procedures. Unfortunately, even science is not cut and dried. When new information is discovered, the theory explaining the finding can be subject to many interpretations. Thus, even science suffers to some degree from social pressure when exploring the truth.

The exponential explosion in the amount of scientific information has required scientists to confine their exploration to narrower portions of their field of study. This narrowing of study tends to exaggerate the negative impact of peer review. This problem is made worse by the fact that it has become expensive to do science today.

In order for scientists to do any work at all, they must secure funding for their work. An important requirement for the advancement of scientific investigation is the improvement in tools to make observations. Often the cost of the tools of research far exceeds scientists' salaries. Society often holds the purse strings for funding for scientific investigation, scientists, and their tools. The particle accelerator that was going to be built in Texas at a cost of billions of dollars was eventually canceled, because the federal government, which was the only organization that could fund such a project, had higher priorities for the funds. The space program in the sixties is another example. In the sixties, the United States had established a national goal of being the first country to put a man on the moon as a way of establishing our national superiority over the U.S.S.R. in the field of space exploration. It required a national commitment to justify the funding for the space program. Thus, science, which would appear to be objective, can also be subject to social pressures.

Male Versus Female Viewpoint on Sex and Relationships

Biologists tell us that the primary objective of the male in most species is to mate with as many females as possible. This is true because the male's primary interest is the continuation of his genes. The male has the best chance of achieving this objective by mating with as many females as possible. I believe this may also be a secondary consideration with social groups with low living standards. Large families also ensure the continuation of the male's genes under conditions where the chances of all males growing to maturity are less likely than in more well off societies.

On the other hand, the primary objective of the female is to nurture her young. With this objective in mind, the female seeks to mate with the male who presents the greatest potential of providing the means of ensuring that her offspring will grow to maturity. This desire is so strong that females will usually select a male with less desirable physical characteristics if the male shows a greater potential to be a better or more reliable provider.

With women's liberation and more equal treatment of women, these factors are less important than they once were, but they still have some impact on how people look at the meaning of life and the process of setting goals and objectives. While it is society's objective to provide equal opportunity to women, it is far from certain that women should pursue their goals the same way that men pursue them. Society may be better off if women pursue different goals than men, if these goals have less harmful impact on society and the environment.

I also believe that we have a tendency to downplay the desire of males to be nurturers in the broadest sense of the term. Men from all walks of life today express a desire to spend more time with their family as opposed to putting more time in on the job. With the rapid changes in our economy and globalization of business, the corporation is providing less sense of security than it did in the past. People are realizing that the family still provides a sense of continuity and security.

Emotions and Instincts

While humans consider themselves to be reasoning beings, emotions and instincts can still play a major role in our concepts of what makes a meaningful life. In general, as social animals, we tend to deny the fact that we are in fact animals, with basic biological needs such as eating and reproduction. We wear clothes to display social status as much as for protection from the elements. We have made the act of eating a social activity. Thanksgiving dinner is probably the best example of this in the United States. We engage in sexual activity in private. We accept as a given that public sexual activity would be shameful. Even public nakedness is illegal. We use soaps and deodorants to eliminate any animal odors. The use of soaps and deodorants has become so ingrained in American culture that we automatically react to body odors as being offensive. However, in other cultures some of the smells that we think are offensive are acceptable.

It is obvious from the presentation of the material in this book that information is provided from the viewpoint of science or, in other words, reason. But an important part of the meaning of life is in emotions, which often do not involve reason or go contrary to reason. Love, hate, satisfaction, joy, despair, and happiness are all emotions that are part of the meaning of life. As conscious beings, all of our experience is colored by emotions. The exact same experience for one person could be an entirely different experience for the same person with different emotions. In recent years, science has studied how to impact emotions and, in some cases, has been successful at doing so. The most notable example is the use of Prozac to treat depression.

In the United States, more than in other societies, sex is linked with emotion; especially, when sex eventually leads to marriage. In other societies where arranged marriages are still common, marriage and sex is associated with social status and property. In other words, sex for pleasure and emotion may be separated from sex for inheritance and maintaining genetics.

Science at least has a theory of why we fall in love with another person. According to this theory, we establish certain patterns in our brains as a result of interaction with certain people that make us comfortable with that type of person. If we have established patterns in our brain for tall and dark males or blonde females, we will be attracted to these people in spite of other negative characteristics. In the broadest sense, this pattern building may explain the old saying that we marry our mothers or fathers, provided we had a good relationship with them when we were growing up.

Love is an important part of almost all religious teaching, and it seems to be a basic human need. Love is the giving up of oneself for another person. Thus, it has to have a place in the meaning of life. It has been said that it is hard to explain love, but you will know it when you see it. I believe this comment applies to true love and not just infatuation.

Love is one of those things that cannot be explained, and yet love and shared experience are important to almost everyone and, thus, to the meaning of life. Some people say that love makes them whole. In the movie *Jerry McGuire,* one of the most notable lines was when McGuire said, "You complete me."

Relationships – Yes, Marriage – Not Necessarily

When people are surveyed about the things that are most important to a meaningful life, relationships are always ranked number one. In addition, the percentage of all people surveyed who rank relationships as very important is consistently in the 90 percent range. While a relationship with a sexual partner is important, people also rank relationships with people in general to be important. For most people, these nonsexual relationships are still more important than many other things, such as having a lot of money.

There is definitely a trend away from marriage in general as well as a trend away from long-term marriages. However, this does not indicate a decline in the need for relationships. Rather, it appears that there is a decline in the institution of marriage. Many people, especially people living in northern European countries, have chosen to live in a domestic relationship and have children without getting married. The decline in the influence of religion can account for part of this trend, but a general increase in the desire to be free from governmental or societal controls also accounts for this trend.

The high divorce rate in the United States may be viewed as an indication of a decline in family life or family values. However, it appears that the institution of marriage is suffering from several trends. The emphasis on personal meaning and the need to have more meaning in one's life is encouraging people to make changes in their life when their present marital situation appears to be standing in the way of a meaningful life. Critics may point out that the increase in the divorce rate is due to a decline in responsibility, which is most likely true to some extent. But, an increased emphasis on the self is also an important factor. People have decided that it is better not to continue in a relationship with abuse or frustration. For some women, who have chosen to excel in business or a profession, this choice has also required them to forgo having children. Without children, many couples do not see the need for the formal relationship of marriage. This is especially true for women who do not need the financial security that they once required when they depended on men for their livelihood.

Even today, there is a wide range of attitudes toward the contribution that a loving relationship and marriage play in the meaning of life. In western culture, the purpose of the family is intimacy and self-expression, while in other cultures it serves economic or political purposes.

Also over the course of a sexual relationship, the character of the relationship changes. In the beginning, the relationship is governed more by passion where the physical aspects of the relationship are important. As times goes by, a relationship involves more shared experiences and beliefs. As a result of this change, intimacy becomes more important.

Relationships also affect meaning through the impact they have on self-worth and security. Self-worth is tied to being the one chosen to be in a relationship. An act of infidelity by one partner in a relationship is usually a greater blow to the other partner's self-worth than to his or her sense of trust and security.

The Myths and True Meaning of Relationships

The idea of marriage as a goal in and of itself is subject to the problems of fulfillment that people suffer with all goals. The achievement of the goal will result in a sense of loss. This problem is compounded with relationships that involve passionate love, since most people agree that passionate love is an irrational state of mind. Most people have a false idea that families are suppose to be filled with love. However, they are units that must face many of the largest challenges of economic life. These economic stresses and the stresses that result from children's efforts to establish their own identity can lead to many familial conflicts that may get in the way of love.

In its need to support reproduction, society emphasizes all the positive aspects of raising children, while downplaying the negative aspects. Many aspects of raising children are unpleasant. The task now lasts twenty years or more, and it involves a significant financial burden without any possible financial payback in old age. Parenthood presents one example where a meaningful life may not always involve happiness. Most surveys show that people who do not have children report that their lives are characterized by factors that would indicate happiness, but they also feel that their lives are less meaningful without children. Children can make life meaningful because they provide purpose. Raising children involves many short-term and long-term goals. In addition, the positive status that society places on parenthood provides a sense of self-worth.

Parenthood is one option for meaning that is available to almost everyone. People who have not had an opportunity to get an education or have not been able to succeed in some work-related area can always look to the family for meaning. Even infertile couples have the options of adoption. For people who have enjoyed some success in other areas of their lives, children can provide the most permanent sense of meaning. As we mentioned in the chapter on work, our accomplishments in work are often subject to revision by the people that follow us after we leave, while our accomplishments in raising healthy, well-adjusted children can be passed on for generations.

In chapter seven, we talked about the work ethic, which implies that there are intrinsic rewards to work. We also showed that most of society has rejected the work ethic in favor of the extrinsic rewards that one can get from work such as money. It appears that society has done a better job of maintaining a belief in the intrinsic rewards of parenthood. This is especially true in light of the increasing extrinsic costs of parenthood. Today, most people are faced with the prospect of spending more than $100,000 for each child's college education. The changes in society's attitudes toward these two values and the average person's acceptance of these values shows that even with regard to personal meaning we are unconsciously influenced by the higher meaning of life presented to us by society.

Chapter Nine

The Individual and Society: Meaning and Values

"The nineteenth century was about economic freedom. The twentieth century was about political freedom. This century will be about Americans deciding for themselves what's moral and what's not."
— Alan Wolfe

What Is Our Place in the Cosmos?

Part of the meaning that we find in life is defined by how we as individuals relate to society and the cosmos. There are two scenarios that we as individuals could use to illustrate our place in the cosmos. In the first scenario, the whole universe is a stage play, and each of us is an actor in the play whose role is set by the script. In the second scenario, we are an audience of one, and all the other people in the world are actors in a play for our benefit. In the second scenario, each person has some control over his or her destiny, but still each person does not have control over the entire script.

Some New Age thinkers believe that there is a third scenario. In scenario three, we are an audience of one, but we have control over the entire script in the sense that we create our own reality. We will not examine this scenario in this book. However, I mention it to acknowledge that there are differing views on this topic. It has been said, if you think about a world view as a religion, we need more than five billion religions—or one religion for every person.

Scenario One: The Fatalistic Approach

The first scenario, which is the most common perception since the beginning of recorded history, is that we as humans have little control over our environment. We are completely subject to the forces of nature. This approach, which I would call the fatalistic approach to nature, implies that we can never expect to have any lasting impact on the quality of our life. We are completely at the mercy of nature.

While ancient humans may have believed that they did not have control over nature, written records and artifacts from before recorded history indicate that humans have believed that something did have control over nature. Polytheistic cultures believed that each of the major forces of nature was controlled by a different god. One god controlled the movement of the sun; another god controlled the weather; yet another controlled the growth of crops and fertility in women.

Regardless of the specific beliefs about which god controlled which force in nature, a common belief was that these gods could be influenced by human actions to limit the degree of chaos so that the we would have the basics required to survive. While most religions today now believe in one God, the concept that certain human behaviors will result in reward or punishment is still a widely held belief.

An essential assumption of these beliefs is a superior/subordinate relationship between God and humankind. In a polytheistic culture, a god may only have power over one of the forces of nature—and indirect power over man through that force of nature. In monotheistic cultures, one God is all powerful. Thus, a single God has direct control over our entire life.

Over the centuries, we have become more sophisticated in our knowledge of the forces of nature, but we are still faced with the ultimate questions: Does some power have control over the forces of nature, and does this higher power require that we behave in some particular fashion in relation to this higher authority? As we have discussed in the chapter on religion, certain people in society have been able to assume some of this higher power for themselves. By assuming some of this power, they have been able to assume some control over our behavior. Ultimately, this has an impact on what we think about the meaning of our life.

One of the quandaries that has faced philosophers over the ages is how much control does man have over his own behavior. Are humans in the final analysis driven by the same forces of nature as every other living thing, or are we unique in some way? This dilemma focuses on whether or not we have free will. If we have free will, then we can be held accountable for our actions. If we do not have free will, we cannot be held accountable for our actions in the long run.

As a practical matter, in modern society men and women are held accountable for their actions. This accountability is a part of the implied contract between every individual and society, and the price of living in a particular society and enjoying the benefits of society is to live by the rules of that society. Over the centuries people have had various degrees of freedom to choose in which society they want to live. For the last four hundred years, a significant number of people have chosen to come to America to enjoy certain freedoms and lifestyles. Although it is much more limited, people still move from country to country today.

Scenario Two: Limited Control

In the second scenario, we perceive our place in the cosmos is the way that every child perceives the world. From birth, a child cannot make a distinction between itself and the world around it. To a baby, the world and its body are one and the same thing. At about the age of two, a child develops an identity that is separate from everything else. The behavior, often referred to as the terrible twos, is that period when a child begins to assert its ability to act independently.

At two, children still do not have a sense of right and wrong, but they have a sense that they have some control over their actions. As we grow up, we all develop a self-identity, but I believe that as a result of our socialization, we never really develop a true sense of how separate we are from everything else in the world.

We are fortunate to live in a country that is based on principles that respect the rights and autonomy of the individual. Others have pointed out that, upon close examination, the Bill of Rights, which is a part of the Declaration of Independence, does not truly describe the rights of individuals, but rather the limitations on government to restrict individual activity. In the United States, individuals concede certain powers to the government for the common good and retain all others for themselves. As it has often been said, government in the United States is a "government by the people for the people." This is truly an innovation in government. All other governments have exercised control of people, and any rights given to individuals were conceded by the government to the people.

The above comments on the U.S. system of government were not intended as a lesson in political science, but rather to provide a frame of reference for how we could view our conscious experience of the cosmos. The great wonder of conscious experience, which I believe most people do not fully appreciate, is our concept of self. Whatever we may call it, a soul or a spirit, it is that intangible point in space and in time by which we experience everything else. The unique thing about the spirit or self is that it is not made any less important by the number of other conscious beings in the universe nor by the physical size or age of the universe.

We all have a conscious experience of the universe that is unique. We can assume that, since you and I are both humans, we should be having the same experience, but at best we can only assume that we are having a common experience. Naturally, the one thing that must by definition be different is the point of reference. I am me, and you are you. If the meaning of life is experience, then this unique point of reference is important to that experience.

The Mainspring of Human Progress

We live in a country in which the freedom of the individual is taken for granted as well as guaranteed by the law. Unfortunately, we only recently provided by law these freedoms to all people regardless of race, gender, or religion. The U.S. system of government is one of the few governments to be established in which the rights of the individual was the primary focus from the very beginning.

Our form of government is a relatively new form of government, which has allowed us to learn from the mistakes or misuse of other forms of government that went before ours. We developed a government that is a servant not a master. The purpose of this book is not to analyze why our system of government is better than another, but it is important to point out that our economy has been able to thrive because it provides an environment in which people can thrive.

People thrive in this country, because they have a freedom of choice and enjoy the fruits of their efforts. There will always be some inequities, but on average, our system provides an environment that allows for self-determination. The freedom of self-determination includes not only what we want to do but how we spend our money. The free enterprise system restricts us only to protect the interest of the individual consumer. This freedom is as important as our freedom in other areas.

When talking with some people about the meaning of life, they somehow assume that a search for meaning involves a rejection of material and commercial things in favor of basic values that include religious and family values. This is based on the assumption that only religious and family values are important. On the contrary, freedom of choice, including material and commercial considerations, is an essential part of the meaning of life.

The Final Freedom

Anyone who has ever driven on an interstate highway in the United States knows that people in the U.S. want to decide what is good for them. Most people drive five or ten miles per hour over the posted speed limit.

In spite of the fact that people are aware that highway engineers set the limits at a safe level, and in spite of the fact that they risk getting a speeding ticket, which will result in a fine and increased insurance rates, people still exceed the speed limit. Whether or not people are actually at additional risk from speeding, they don't think that the extra five or ten miles per hour puts them at risk. They decide what speed is best for them.

In recent years, people have taken the same approach toward setting a meaning for their life and the moral values that they will use to guide their daily actions. In the past, people set goals and relied on moral rules that were handed down from a supreme being or at least some higher authority than themselves. Today we make moral decisions based on what other people require and what are the consequences of our actions.

While we have enjoyed economic and political freedom for many years, it has been only in recent years, generally since the sixties or seventies, that most people have felt enough autonomy to practice moral freedom. For people who are old enough to remember the sixties, this was a decade in which established authority was challenged. Americans had civil rights and the Vietnam War on their minds. They questioned whether or not the United States should be involved in the war. They didn't trust the government's often unclear and misleading information on how the war was being conducted. And it became evident in the sixties, that established authority was sometimes morally wrong. Many of the best political thinkers, such as John Locke and Immanuel Kant, believed that if people were allowed economic and political freedom, moral authority was still required to provide a frame of reference. However, the needs of society have changed since the time of Locke and Kant.

We Live in a More Complex Society

Besides the major moral issues of the time, such as civil rights and foreign policy, our society has become so complex and is changing so rapidly that is difficult for any hard and fast rules to be appropriate in all situations. Self-discipline is a moral value that everyone would agree is a good thing. However, the application of such a principle in different circumstances can have different outcomes. Some outcomes may be viewed as good, and other outcomes may be perceived as bad.

Many people point to our level of affluence and criticize young people today as not having the discipline to work for the things they want. On the other hand, we have people who are workaholics. People striving to get ahead in business will sacrifice the emotional needs of their family and possibly their own health for their job. Ironically, these same people, who could be considered to be loyal to their employer, may find themselves without a job when their company is acquired by another company and the acquiring company lays off half of its workforce. Thirty years ago, the person who was loyal to a company could expect loyalty in return.

There are many nuances to the practice of honesty. Today people can be honest, but they may also have to be politically correct. People can also be brutally honest in the sense that the honesty inflicts unnecessary harm on the person that is the subject of the honesty. Practically no one will refrain from lying to a telemarketer to get him or her off the phone at the dinner hour.

Morality is the product of social evolution. It is a part of the social contract that the human species has developed for the survival and improved quality of the life for the species. The primary objective of social evolution is the common good of the species, which in turn provides in general for the individual that takes part in that social contract. As a means of refining and defining the common good, we have to define certain moral values that support the primary objective of the common good. As society becomes complex, our interactions with one another become more complex. Our superior intelligence allows us as humans to use that intelligence to adapt our responses to situations as we anticipate the consequences of our actions. Thus, we are now faced with the need to decide for ourselves what the true moral outcome is of our actions, although that at first may appear contrary to established moral values.

Since the availability of reliable birth control medications, Roman Catholics have been confronted with the conflict of following the church's teaching on birth control and incur the significant additional financial burden of raising a large family or ignore the church's laws and have families they can financially manage. This dilemma has led to a another moral issue. In raising a large family, do Catholics have a responsibility to provide equally for all children? What happens when they can only financially provide higher education for some of their children and not all of them? This conflict is an example of many more conflicts that all people will face in the future, as technology provides us with more options.

Americans tend to distrust elected officials, especially those that claim that their authority comes from a higher authority. This country was founded by people who fled from Europe where they either suffered under leaders who claimed their authority from divine right or were involved in religious wars that were based on conflicting values.

While people are using more judgment in determining what is a moral act, the process still often involves looking to the guidance some higher value may provide. It is not surprising that with increased communications and a global society that people will look to a wider range of authorities. In fact, it will become essential that people who interact with each other have some common ground with regard to moral value and life's meaning.

While the extremists that attacked the World Trade Center may never find acceptance in a wider society, the West will have to develop a greater appreciation of the values and goals of the Islamic world if we are to eliminate such acts of terrorism in the future. It is ironic that some of the values of self-determination that originally were taught by Mohammed are the same values that contributed to the success of the United States. As some people have characterized it, the Islamic religion has been hijacked by many of the authoritative governments in the Moslem world for their own political benefit. It is worth noting that even communism was never practiced according to its original principles, because it too was hijacked by the government of Russia for its own goals.

The facts of history are that Mohammed was a retired businessman living in a prosperous region. Europe, at that time, the sixth century, had fallen into the Dark Ages. Mohammed believed that God judges men but he does not control men. He also believed each individual is self-controlling, and each person is responsible for his or her own actions. This is a far cry from the tenets espoused by fanatical Islamic groups today.

Regardless of whether we adopt moral values as our own because they are handed down to us from a higher authority or we develop them on our own, there is a process of providing the motivation or rationalization to following these values or setting these goals. There are stories of people who have suffered unbelievable torture and degradation in prison camps during the World War II and Vietnam. These people relate that the only way that they were able to hold onto hope was to believe that they would eventually be free, or to find some good in what they were enduring. Even though these people were in prison, they accepted their fate not because they had no other choice, but because they had decided to accept the situation on their own terms. This acceptance or integration of values and goals is essential to living life with true meaning and satisfaction.

Some of the thoughts in this section echo the findings of Alan Wolfe who conducted interviews with several people that he believes represent a cross section of American society today. These findings were summarized in an article that appeared in the *New York Times Magazine* on March 18, 2001. A more detailed account of the survey was published in April 2001, in a book entitled *Moral Freedom: The Search for Virtue in a World of Choice.*.

Step Outside of Life

When we attend a play, we sit in the audience outside of the action and watch real people act out a story that has been set in advance in a script. Even if we have seen this play many times and know how it will end, we still enjoy watching the story unfold.

For me personally, life is like a play in which I sit *inside* the action. Everyone else in my life plays a role in my life. No matter how rich or famous some people may be or how unfortunate other people may be, they all have a role on the stage of my life. Because we live in a complex society, there are many thousands if not millions of other people who, though unseen by me, play a role in my life. All you have to do is experience a storm that cuts vital services such as electricity or telephone to see how many people have a direct impact on your daily life.

If you look at life simply in economic terms, one might say that the rewards are not distributed equally in the sense that some people receive more money for the time they spend working. However, in reality, the rewards we receive for our work is not the money we receive but the goods and services that we purchase with this money. Most of the goods and services that we purchase today would not be possible without the coordinated efforts of many people. The millionaire as well as the factory worker can watch a football game on television. They both can afford the service of a reliable electric company. They can both afford to buy a television, which represents only a few days' pay even for the factory worker. Even though the football players may be earning salaries in the millions, both the factory worker and the millionaire can watch the game for the cost of electricity.

It sometimes helps to look at life as an observer rather than a participant to understand the meaning of life. The meaning can be more insightful if we can step away from the biases that we have learned over the years. Unfortunately, it is hard to separate ourselves from these biases. Jimmy Carter, who has developed a reputation as a skillful negotiator in international crises, was ridiculed for consulting his daughter Amy, during his presidency, about pressing issues of our time, such as nuclear war. However, children who have not yet developed political agendas are often accurate barometers of the cultural climate.

After the September 11 terrorist attacks, a reporter asked children what it meant to be an American. One ten-year-old girl replied, "It means we love freedom and peace, even if we have to kill people for it."

Another Place and Time

Sometimes when I think about how people lived two hundred or three hundred years ago, I think that times must have been hard. However, if I think about how people lived one thousand years ago, I would image that life must have been even harder then. On the other hand, many people look back at how people lived only fifty years ago and wish for the days when life was less complicated. At the rate that technology is improving, I am sure that people one hundred years from now will think that our lives today were hard. In general, I don't think the times we live in have much impact on the meaning of life or even the overall quality of life. In the past, people may have had to endure more physical discomfort, but life was still meaningful.

A recent article in the *New York Times Magazine,* written by Ron Suskind and titled "A Plunge into the Present," examined the last twenty-five years of life in a society, the Ibaten, which lives on a volcanic island near the Philippines. Until 1977, when two missionaries arrived, this society lived a primitive existence. Since then, the Ibaten people have been introduced to monotheism and free enterprise. In some ways, these people have gone through several thousand years of progress in just a quarter of a century.

However, in other ways, they are more like a society looking in on the twenty-first century. They have satellite television, which allows them to watch world events, including the World Trade Center attack, as they occur. But, they are located on an island, Babuyan, which is off the usual trade routes. Travel to and from the island is difficult, and getting goods to the island is even more difficult. I believe it is fair to say that the Ibaten have some ability to be selective about which of our modern ways and goods they would like to make their own. As Suskind writes, the island provides an opportunity to examine "the very idea of modernity and what constitutes human progress."

Suskind describes how building a basic water system had an impact on Ibaten society, because before the water system, trips to the spring offered people an opportunity to socialize. Just like people gather around the water cooler in the office. The Ibaten were also taught a nondenominational Christian religion, which provided the people with a new sense of self-worth and a sense of individual destiny. The missionaries sent several of the young islanders to college on the mainland. When these islanders returned from the mainland, they became the beginning of a quasi upper class of the Ibaten. These college-educated people were able to earn $400 a month, a large sum for the island, as teachers for the Philippine Department of Education.

Islanders chose different ways to participate in the progress that the island enjoyed. Some people on the island started a business of supplying fresh fruit and water to passing fishing boats. One couple was able to open a cooperative store to sell basic supplies, after the missionaries taught the wife basic accounting. Later the store bought a generator and a diesel-powered refrigerator, which allowed the proprietors to sell electricity and ice to other islanders. The article details that as the Ibaten make progress, their ambitions for more progress becomes greater, which outpaces their capacity to make these plans materialize. But it is important to maintain a sense of forward progress or hope collapses.

Suskind ends the story with an account of a meeting with islanders. The people review some of the progress they have made and list some projects for the future. Everyone agrees that a high school is needed, but it is way beyond their means at this time. A more practical project is to build a cement bridge over a river that divides the island during the monsoon season. However, to build the bridge, they need steel reinforcing bars, and there is no boat on the island big enough to bring them to the island. Furthermore, hiring a boat specifically to transport the reinforcing bars would be too expensive.

The solution that they devise to this problem illustrates social progress in action. Just before this meeting, the island was hit by a typhoon. In the aftermath, some of the islanders were being forced to sell some of their livestock. They plan to negotiate with the livestock brokers, who have large boats, to bring the steel bars, when they come to buy the island livestock. In addition, the islanders who must sell their livestock, will work together to get a better overall price for the livestock.

This story illustrates that while each of us establishes our own meaning in life and then lives our life based on the meaning we have set for ourselves, the world we are exposed to and the opportunities that the world presents us influence our actions. It also appears that the greater range of opportunities offered by a more economically and technologically advanced society results in a more diverse society. For whatever reason they do it, some people decide to make the extra effort. When people make an extra effort, it is not completely in their immediate self-interest. However, some acts of altruism, such as providing services at little or no cost, are good for society in the long run.

Meaning and Values

Society has been able to associate morality with meaning, in the sense that behavior that results in the greatest benefit for society is the most meaningful for the individual. Biologists tell us that animals are subject to conditioning. In other words, animals will continue activities that provide positive feedback and will discontinue activities that incur negative feedback. This is part of any learning process. As conscious beings, we have projected this behavior to some higher power if it exists. We assume that if we engage in some behavior that is satisfying to that higher power, we in turn will be rewarded by this higher power, because our behavior is appealing to this higher power. In a sense, we have projected our reaction to meaningful activity to some higher power. However, we can have a meaningful life without a higher power, at least a higher power that is in control of the universe. We can have a meaningful life simply in the context of the society in which we live.

Our life only has meaning in relation to the world around us and our place in society. In one sense, we are all selfish beings in that we have free will and we do what we think is best for us. However, as a result of social evolution, we have come to understand that what is good for society is also good for us. At the same time, society has learned that an environment that allows the individual to enjoy the benefits of his or her efforts makes the fastest progress, at least in terms of physical comfort and quality of life.

As society becomes more complex, our value systems must become at the same time more complex and flexible. Also as groups become more interdependent, they must share some basic common purpose and meaning. Paradoxically, while we must all share some common purpose, we all have an experience of reality that is truly unique.

Although the Ibaten people were introduced to modern society by missionaries, they seem to exhibit behaviors that are universal to the human species. It appears that they freely engaged in activities that we in advanced societies believe makes life more meaningful. Thus, one could say that there are some values and activities that are universally important to a meaningful life. Also, each of us goes about this process in a unique way.

Chapter Ten

Guidebooks to Happiness

"I believe the very purpose of our life is to seek happiness. That is clear. Whether one believes in religion or not, whether one believes in this religion or that religion, we all are seeking something better in life. So I think, the very motion of our life is toward happiness."
— His Holiness the Dalai Lama

An American businessman was at the pier of a small coastal Mexican village, when a small boat with just one fisherman docked. Inside the small boat were several large yellowfin tuna. The American complimented the Mexican on the quality of his fish and asked how long it took to catch them.

The Mexican replied, "Only a little while, señor."

The American then asked, "Why didn't you stay out longer and catch more fish?"

The Mexican said that he had enough to support his family's immediate needs.

The American then asked, "But what do you do with the rest of your time?"

The Mexican fisherman then said, "I sleep late, fish a little, play with my children, take a siesta with my wife Maria, and stroll into the village each evening, where I sip wine and play guitar with my amigos. I have a full and busy life, señor."

The American scoffed, "I am a Harvard MBA and could help you. You should spend more time fishing and with the proceeds buy a bigger boat. With the proceeds of the bigger boat, you could buy several boats. Instead of selling your catch to a middleman, you could sell directly to the processor, eventually opening your own cannery. You could control product, processing, and distribution.

"You would need to leave this small coastal fishing village and move to Mexico City, then LA, and eventually New York City, where you will run your expanding enterprise."

The Mexican fisherman asked, "But, señor, how long will all this take?"

To which the American replied, "Fifteen to twenty years."

"But what then, señor?"

The American laughed and said, "That's the best part. When the time is right, you would announce an IPO and sell your company stock to the public and become a rich man. You would have millions."

"Millions, señor? Then what?"

The American said, "Then you would retire. Move to a small coastal fishing village, where you could sleep late, fish a little, take a siesta with your wife, and stroll to the village in the evenings where you could sip wine and play your guitar with your amigos."

The Process of Living

This story is full of insights, but the one that hits home for me is that it is easy to get over-involved in the process used to get to an objective. In some cases, as in the story above, we can lose sight of the objective altogether. In this book, we have looked at how science can give us a different perspective on truths that are essential to the meaning of life. The social sciences, such as psychology and sociology, address the process of living itself. Many of the self-help books published today provide information about human behavior that has been learned by the social sciences.

A major objective of this book was to get you to take a step back from some of the basic concepts about the meaning of life and maybe come away with some new ideas. On the other hand, many other people have written many great books on topics that relate to the meaning of life, even though they may not address head-on some of the issues that I have covered in this book. While I caution the reader to be careful not to get caught up in the *process* of living a meaningful life, I believe that some process is helpful. In this chapter, I provide an overview of both concepts and process provided by others.

The self-help section is one of the largest nonfiction sections in most bookstores. There are a large number of titles on the subject of a meaningful life, and each title takes a little different approach to the subject. However, there are common themes in many of these works.

One common theme is the idea that life is more a journey than a destination. One could assume, if life is an experience rather than a constant effort toward specific goals, that planning and process does not enhance or contribute to a meaningful life. However, that is not true. You may have read one or more of the books covered in this chapter, since they have all been bestsellers. I am sure you will recognize the self-help books in the bibliography. I recommend reading every one of them from cover to cover.

Confusing Happiness with Pleasure

According to the Dali Lama, many psychologists, and a number of self-help gurus, the purpose of life is to seek happiness. Often when people question the meaning of life, they are asking why aren't they happy. Psychology was one of the first sciences to directly address some of the same issues that once were only the purview of religion and philosophy. Today, some people talk to their priest or rabbi to get some of the same counseling that other people get from a psychologist or social worker.

Psychology and psychiatry are still confronted, in one way or another, with the longstanding problem of the separation of the mind and the body. Before modern science, many mental illnesses that we treat today with medicine were once believed to be problems with a person's soul. Some mental illness was viewed as possession by the devil. Sigmund Freud, considered by some to be the father of psychiatry, developed his principles based on our conscious experience, with little connection to what might be going on inside our bodies.

Today, most of psychiatry seeks to address problems relating to our conscious experience in terms of chemical processes that go on inside our brains and bodies. Many people now look to a pill or herbal remedies to make them feel better. Although psychology and psychiatry cover a wide range of dysfunction, the key aspect of these disciplines is to improve people's conscious experience.

According to the Dalai Lama, we have to train our minds to identify those things that lead to happiness and those things that lead to suffering. The source of our happiness is our state of mind or our outlook on life and the people around us, and what our self-worth is associated with. If we establish our self-worth in terms of material things or success in our job, we will only be happy for as long as our material possessions are available or our success in work continues.

Many people confuse happiness with pleasure. The problem with pleasure is that it only lasts as long as the thing that gives us physical pleasure lasts. Satisfying the desire for the physical thing ends the pleasure.

We have to learn to be happy. Being happy is a process of training the mind over time. All things that lead to lasting happiness take time and effort. We have to learn how positive and negative emotions affect us. We learn things on many levels. Therefore, we may learn some things at one level in a short period of time, but it will take longer to learn others at a deeper level. The most important aspect of the training is self-discipline.

Many people may think that self-interest is all bad. We could get this idea because we live as predator animals. However, altruism, which is a form of cooperation, is also in our self-interest, because we have learned that we stand a better chance of survival if we cooperate with others. This behavior is characterized as compassion. At the same time that we understand how we are interconnected with other people, we also have to understand the importance of self-reliance. However, our self-reliance must be tempered by the knowledge that in a complex society such as the one in which we live, we depend on many people for almost all of our basic necessities.

We also have a need for intimacy with another person. However, this intimacy does not necessarily have to be with a spouse or a significant other. While we have a need for intimacy, we must temper this need so that we do not become dependent on others. In order to develop sympathy, we have to develop empathy with others. In order to develop empathy, we have to be able to deal with people in terms that they can understand.

While a hunter may not be able to understand the suffering of the animals he kills, he can understand the suffering of his dog if it gets caught in a trap. To have empathy, we also have to understand a person's background or what is going on in his or her life. Some people may appear remote or reserved. This reaction may be more the result of other things in their life rather than the immediate circumstances of an encounter. Empathy also requires us to have an open mind and be honest with others.

To determine the source of conflicts in relationships, look at what the relationship is based on. If a marriage is based on sex or romance, the marriage could be in trouble if the original circumstances change. Also if a friendship is based on wealth or social status, that relationship will also be in danger if the circumstances change. Compassion is based on the things we have in common as human beings, and we must we able to accept another's suffering. If we show compassion, we will receive compassion in return.

No Pain, No Gain

Western society has been able to reduce the suffering that we endure as a result of the natural elements. Thus, we have come to believe that we can eliminate all suffering, but suffering is still a part of existence and we must accept this fact. Most good self-help books say that suffering is a part of life. If we learn acceptance of suffering, it is easier to bear. On the other hand, if we feel that in some way our suffering is due to something that is unfair, our suffering is only increased. Perspective is more important than reality in the case of suffering. We may think that we are the only person that is suffering, but when we realize that other people suffer also, it seems to be easier to bear.

It is important to remember that good things come after suffering. In the gym, we use the expression "no pain, no gain," but it can apply to other situations in our lives as well. Physical pain is essential to our learning. We discover what is good or bad for us through pain; touching a hot stove is often used as an example. Pain sensation also helps us with our self-awareness of our bodies.

Pain is also part of the process of change on both the personal level and for society as a whole. Often major life-changing events involve a great deal of pain. Pain can create a sense of urgency in our efforts to solve a problem.

For change to be significant, it must be learned by habit. In the ultimate, change becomes a part of our genetic makeup. Major change comes in many small steps. Since change can take place in many small steps over a long period of time, it is important to develop a wider perspective that will allow us to focus on the long-term objectives. Realistic expectations when setting goals are important.

Training the mind to be happy involves a process of replacing negative and painful emotions with positive ones. Western society tends to use psychotherapy to address our neuroses rather than looking at a more complete or holistic way of thinking as is done in the East. Some people think that anger and hatred are a natural part of our mind. Thus, we do not expect to be able to get anger and hatred out of our nature. But through training, we can replace anger and hatred in our minds with love, compassion, and forgiveness.

The idea of replacing negative emotions with positive emotions is based on the concept that negative emotions have no valid foundation, whereas positive emotions do have a valid foundation. For example, we can realize that no one wants to suffer. Therefore, we can develop compassion from this realization. On the other hand, everyone has a need to satisfy basic needs, but excessive desire turns into greed, and greed has no basis in the basic needs for life.

Buddhism teaches that when we understand the true nature of reality, we can also understand why negative emotions have no basis in reality, while the positive emotions do. The West tends to accept that the positive emotions are good based on faith, while the East bases its ideas on reasoning and training of the mind.

Eastern thinking believes that negative or afflictive emotions can be offset by positive emotions. When people are depressed, they think that all things are hopeless. One failure can result in a complete destruction of self-worth. By showing a person that there are some good things in life and that they have positive capabilities, the negative emotions can be offset and rendered less destructive.

Anger and hatred will destroy us. So we must replace self-hatred with self-love. We have to offset anger with patience and tolerance. We are anxious because we fear a loss of self-esteem and looking foolish. When anger arises, we should stop and analyze what factors caused our anger and decide whether the anger is constructive or destructive and whether the anger is an appropriate response.

Scientific study has shown that anger creates a state of physiological arousal that can lead to more anger. If left uncontrolled, anger tends to escalate. If anger is offset with patience and tolerance, we also protect ourselves from the long-term physical effects of unrestrained anger.

Some people may think that responding to hurt with patience and tolerance is a sign of weakness. However, it is a sign of strength, since one must have the strength to remain firm in the face of adverse situations and conditions.

The concept of "sincere motivation" involves the idea that we should not be anxious in any situation that would normally be a source of anxiety. We should assure ourselves that the motives of others for taking actions that are making us anxious have the best intentions. In other words, the closer we can get to altruism as our motivation for actions, the less stress we will have in normal stress-producing situations.

Flow and Control

Some people have studied what it means to have a good experience, which may not be associated with happiness or pleasure. In his book *Flow: The Psychology of Optimal Experience,* Mihaly Csikszentmihalyi effectively starts out where many other books finish. He accepts as a given that we have little choice with regard to many of the major events in our life. For example, we do not get to choose our parents, how good looking we are, or how smart we are. As we go through life, we are subject to the whims of fate that bring us good and bad fortune

In spite of the fact that we live under conditions that could make people despair, he observes that we all enjoy periods when we feel that we are in control of our destiny, when we have a sense of exhilaration. These periods provide the basis of fond memories. These are the times that provide meaning to our life, and if we could, we would make these times as large a proportion of our lives as possible.

According to Csikszentmihalyi, these times are active times when we have used our mental and physical capabilities to their limits. These are times when we are trying to achieve something that exceeds anything we have done before or even charting new ground for the human species, although in most cases we are working toward some goal that is much less grand. He calls this time when we make things happen as an optimal experience.

From accounts of what it felt like to have these experiences, Csikszentmihalyi developed a theory of optimal experience which he called "flow—the state in which people are so involved in an activity that nothing else seems to matter; the experience itself is so enjoyable that people will do it even at great cost, for the sheer sake of doing it."

Csikszentmihalyi observed that all people, men and women, young and old, in many cultures around the world described optimal experience in common terms. Therefore, he decided to gather subjective data about various aspects of optimal experience. His work began about twenty years ago, and many other people are applying flow concepts in schools, business, and clinical psychotherapy.

To me the most interesting thing about the study of flow is the realization that our conscious experience can be the object of subjective study. Eastern religions teach the value of meditation, but meditation is an activity that is done in contrast to, and in another time, than day-to-day experience. We live in a world of social controls that provide rewards to maintain the social order. Much of the frustration of contemporary living comes from the fact that once we acquire these rewards, they prove to be fleeting.

Flow is offered as a way of dealing with this system of social controls. It is not offered as an alternative, but more as a supplement to the goals offered by society. In its simplest sense, flow allows us to get a little more control over our lives. This control can relate to minimizing the impact of negative events as well as maximizing the experience of positive events.

It is also interesting that flow is not offered as a quick fix or something that does not require work. Csikszentmihalyi's book, like many others in the self-help area, points out that it takes time to modify our habits and desires. The author also indicates that the concept flow itself is subject to constant revision in relation to the changing environment.

Life Planning and Mission Statement

Corporations in the last few years have been establishing and communicating mission statements. According to the law, a corporation is a person. Therefore, it may not be as unusual as it first seems to use the corporate mission statement as a model for an individual's life mission statement. Some people may think that a corporate mission statement is just a glorified advertising slogan, but if a mission statement is well prepared, it's the most important communication that a corporation can make.

Although most people think that a corporation is only responsible to its stockholders, in order to be successful, a corporation must interact with and be responsive to a large number of people whom all have different roles and objectives. Included in this list are stockholders, employees, customers, regulators, and the community in which the corporation operates.

Writing a corporate mission statement is difficult because it must be short enough to hold people's attention, while at the same time communicate the essence of why a corporation is in business, what its goals are, and how it plans to accomplish those goals. A company that is serious about its mission statement will have it widely published, and ask its employees to refer to it frequently as they do business.

It has become almost routine to a see a company go out of business or be acquired by a competitor because it lost track of what business it was really in—part of its mission—which in turn put it out of touch with its customers. This eventually resulted in declining sales and profits regardless of how strong its market position might have been at one time.

Individuals need to have life mission statements—what is my purpose in life, how can I accomplish what I want to do, and most important, will I enjoy working toward my goals? Continuing the corporation analogy, a successful corporation grows and makes a profit, a successful individual is happy. With a life mission statement, everything in life will have meaning. Some things may not have a positive meaning, but they will still have meaning.

According to existential philosophers, people can develop either authentic or inauthentic life themes. An authentic theme is a theme that an individual develops based on rational evaluation of experience. It does not matter whether the theme is right or wrong; it only matters what the person genuinely feels about and believes in the theme. An inauthentic theme is a theme that people think must be done because everyone else is doing it—following the crowd, trends, or fads and believing there is no alternative.

There are pros and cons to both kinds of themes, when setting a meaning for life. Inauthentic themes can work well if social themes are basically sound, but they do not work well when the social system has gone bad. The classic example is the bureaucrat the works toward the ends of a corrupt government. An equally good example, which is not cited as often, is the Inquisition in the Catholic Church that took place in the Middle Ages. During the Inquisition, religious principles were perverted for the benefit of a few or a specific group.

Authentic themes can have drawbacks as well. If the theme is contrary to established thought, it may be viewed as crazy or destructive. Because authentic themes are established through a personal process of defining the purpose of life, they may be new and unique.

If we were to take a poll to determine whether a person has an inauthentic or authentic theme for life and ask the same person a few other things about his or her life history, I believe intuitively that we would see the following: People who had a happy or relatively unstressfull childhood develop inauthentic life themes. People who had above average stress in their childhood develop authentic life themes. Often people in the arts attribute their creativity at least in part to stressful early life experiences. At an early age, established life themes were not working for these people and they set their own path.

People who established authentic life themes later in life most likely changed their life themes either as part of a mid-life crisis or in reaction to a major life-changing event, such as a serious illness or the loss of a close friend or relative. Obviously, all life themes that are developed as a result of some major event do not all lead to the same positive or constructive outcome. Some people are changed for the better, and others continue to feel cheated or unfulfilled. The big question is what allows only some people to benefit from their experience.

It appears that some people are able to draw upon the order that other people have found in the apparent chaos in the universe to put order in their own mind. While setting a life theme is a personal process, it does not preclude us from looking at what truths others have been able find over the ages, or to enjoy the order that others have been able to create, such as philosophy, religion, music, art, literature, and, more so today than ever, science.

For example, in recent years psychology and all science in general has struggled with religion. Psychologists have observed the harm that misguided religion can inflict on people, and they are tempted to discard everything that religion has to offer. Religions, like any social organization, are subject to the corruption of the people that are involved with them. Scott Peck, the author of *The Road Less Traveled*, and others have observed that if we discard all religion and what it has to offer, we lose the model for spirituality that can offer us much toward establishing a meaningful life theme.

By the same token, science has rejected the biblical account of creation, but the order that science has been able to uncover in the extreme complexity of nature makes me believe that this order can be a guide in establishing our life theme. Some people believe that we must accept the existence of God in order to have any meaning of life theme. For these people, that is part of their life theme.

However, I believe a meaningful life theme is possible for people who are not sure about the existence of some all-powerful, spiritual entity other than the mere force of nature. Men sometimes think that we are separate from nature, but we are a part of nature. Thus, what we can learn from nature should help in the process of developing a life theme. In general, we should compare our personal experiences with what we can learn from established beliefs, accept what works for us, and reject what does not.

We live in a world where every established system has something to offer us in the process of developing goals, but none seem to have all the answers. Some people by default accept the system that offers the most positives for them, while others are hoping that some new system, possibly science, will emerge to provides all the answers, including not just what our place is in the universe but what our goals should be.

Although I believe that science can provide guides to understanding the meaning of life, I side with those who focus more on the journey than the destination. Science has already revealed some things that are difficult for humans to understand, such as the fact that light has some properties that indicate that it is a particle (matter) while at the same time it has some other properties that indicate that it is a wave (energy). Perhaps life is both a journey and a destination, or it at least gets meaning from both the journey and the destination.

How We Establish Meaning

Although we all establish meaning in life, most people do it without any understanding of the steps that we take in the process. Since the thesis of this book is that individuals should come to a meaning of life on their own terms, it would be helpful to understand the process involved. In this chapter, we refer to the sense of meaning as setting goals, objectives, and purpose.

Psychologists have determined that people go through a series of steps in the process of determining who they are and what they want to do with their lives. Everyone starts with the need to preserve the self and keep their body functioning. The objective is to satisfy basic needs for survival, comfort, and pleasure. Once the self is secure, the individual will adopt the values of the social system in which he or she lives. These can include the family, other larger communities from a tribe to a nation, and religious and ethnic groups. This step involves more complexity of the self in spite of the fact that it requires conformity to community standards. At this point, the main purpose in life is to grow and improve one's capabilities in relation to one's potential. The final step is a turning away from the self and an integration with other people and the adoption of universal values.

In this process, there is an alternating attention to the self and to others. It is my belief that this process is another manifestation of adaptive complex systems, which were discussed in the chapter on uncertainty. As a part of the system, the individual invests energy in the needs of the organism or the goals of the community. However, even this energy is, in a sense, invested in the self because the self receives a return on the investment: it receives the rewards of belonging to the community and sharing in the community's success. The group or community is more successful with the individual than without him or her; the group can achieve more than the sum of what they can achieve when they work alone.

Once an individual achieves a sense of belonging to a larger human system, the individual tries to determine and develop his or her individual potential. At this stage, an individual experiments with different ideas and ways of providing for economic needs, while at the same time achieving some emotional rewards. This can be the time when people have a mid-life crisis, because they may be frustrated by the limits of their capabilities. This is especially true when an individual has established high goals and objectives. In the final stage, the individual redirects energies to a larger system and universal values.

Not everyone moves through all these stages of development. Some people never have the opportunity to move beyond the survival mode. Fortunately, in the survival mode, self-interest is enough to give meaning to life. Most people are comfortably located in the second stage of development where family values and support of a larger community provide a sense of well-being and purpose. Only a few people move to any higher level of purpose and values.

Psychologists tell us that there is no reason why we must go through this process of developing our capabilities and recognizing both our potential and our limitations, but we will feel that we have lived a fuller life if we do.

We live in a complex society, which provides us with many alternatives when it comes to setting goals and choosing values. The number of options is great, and the factors involved in making a choice are numerous. In the last one hundred years, we have gone from one extreme to the other, with regard to choices. Once women were considered property, and today they can consider a life in business, education, or the arts. Or they can choose the traditional roles of wife and mother. The point is to choose goals wisely and be aware that simultaneous goals can be conflicting; we can't do everything and be everything.

Many people today are choosing second careers that are not motivated by a mid-life crisis. Today, it is popular for the successful person to "give something back" as a community volunteer or change careers to something that is less stressful.

Everyone must eventually figure out their ultimate purpose on their own. With the best of intention, parents may try to impose their goals on their children, but this can only lead to frustration. Many institutions and social groups are also offering us potential goals. Aristotle said that the unexamined life is not worth living. Without self-examination, we can easily waste our lives pursuing conflicting goals or goals that are meaningless for us personally. One way of clarifying personal goals is through reflection, and the reflection should be done on a regular and frequent basis. Some people use a religious retreat as a means of regular reflection. Most people take vacations as a change of pace from their everyday lives. Unfortunately, vacations are normally packed with so much activity that there is little time for reflection, but vacations could be used for this activity. Many religious orders reflect on what they did each day to attain goals. Commuters often find the time traveling to and from their jobs as a good time for daily reflection.

Setting Goals

Before putting a great deal of effort into a goal, it is worthwhile to ask some basic questions. Is this something I really want to do? Is the price that others and I will have to pay worth it? Is this something I enjoy doing? Am I going to continue to enjoy doing it? Will I be able to live with myself if I accomplish it? What do I do after I accomplish it?

These questions seem simple enough, and they should be if we practice reflection on a regular basis. In fact, they can become a part of our internal mental process. We can ask these questions almost in our unconscious. If we don't practice reflection regularly, they could be impossible to answer, because we have been so involved in action that we have lost touch with our own feelings.

Ironically, one factor that separates us from other animals, our consciousness, is also the source of much of our suffering from anxiety, frustration, and loneliness. If we can live a life that is directed and focused by goals, we can achieve an inner harmony that can reduce this suffering. The reason that animals other than man seem free from these problems is that most of the time animals' perceptions of what has to be done coincide with what they are prepared to do. When animals are stimulated by hunger or sexual hormones, they find something to eat or find a mate. Animals become frustrated when their biologically programmed goals are not realized. When they are satisfied, they do not suffer from confusion and despair.

Bernard Baruch, the famous financier, once said, "Both men and squirrels put things aside for a time when food may not be available. The difference between men and squirrels is that the squirrel knows when it has enough, man does not." In other words, only man will acquire wealth just for the sake of having wealth, long after any possible material needs would require it.

Choice is a double-edged sword. Ironically, many people today make sacrifices so that their children can get as much education as possible. The primary motivation of providing as much education as possible is the desire to give our children more choices in choosing life goals. By giving our children more options but not preparing them to make these choices so that they do not lead to frustration, we are contributing to our own frustration and cumulatively to the frustration of our society.

Understanding Our Self

The first step toward happiness is an understanding of self. We have to understand our past, because our perceptions of past have an impact on how we see the present regardless of what the true reality may be.

One part of psychology devotes itself to psychotherapy. Through talking with a professional guide, people can examine their past and hopefully come to an unbiased evaluation of that past. Many psychologists say that all people are a little crazy. Only a small part of the population is so dysfunctional that it cannot live in society. A much larger proportion have some dysfunction that prevents them from living a fully rewarding life as could be possible with a little better self-understanding. The biggest impediment to self-understanding is the denial that we use to help avoid the pain that we suffered as a child.

The basis of our personality and the way we interact today with people is an extension of the way that we interacted with our parents when we were children. If as children we lived in a family in which we were mistreated, spoiled, subjected to unreasonable standards, or abandoned by a parent, today as adults we live in a way and relate to others in a manner that may either continue these types of unhealthy relationships or allow us to avoid relationships entirely to keep the same thing from happening to us as adults. Our childhood experiences can result in character flaws such as unreasonable fear, greed, and self-preoccupation.

For example, if we lost a parent through divorce or death when we were a child, we may avoid developing intimate relationships as adults. That doesn't mean we head for the woods and live as a hermit. We can sabotage relationships with far more subtle behavior that may not be obvious and could be easily rationalized by the person with the dysfunction.

Unfortunately, even after some people develop a rational understanding of the past and how it affects their relationships, they do not enjoy the happiness that they expected when they attained their goals. These people are still unhappy because they are too focus on the attainment of things, especially the things they do not have. People who focus on the things they *do* have rather than the things they do not have, no matter how little they have, can be truly happy. People who are not happy with what they have could be said to be too focused on the future. The moral of all this? Don't focus on the past and don't focus on the future—focus on the present.

We cannot always guarantee the results of our actions or that the results that we get are the result of our efforts. Therefore, we should focus on our intentions. If our intentions are right, the outcome will always be right regardless of the actual results. "It's not whether you win or lose, it's how you played the game." Some people spend so much time focusing on the past or the future that they never really have any present. We still must learn from our mistakes in the past, and we must still plan properly for the future, but we must retain some balance. Be excited about life but be comfortable regardless of the outcome.

Strategies for Living

Some books provide a set of rules and a workbook that can be used to set life strategies. The primary focus of these books is to guide us through the process of taking an objective look at the world in which we live and how we approach the task of day-to-day living.

Unfortunately, many people are less than objective when evaluating their personal situation. Thus, many of the problems that we may face in everyday life are a consequence of our failure to take an objective look at our life and to develop a disciplined plan for life.

While some people may say that it doesn't pay to be cynical, it also is important to be practical when considering what we have to do to survive and be successful in our complex world. Some books make it look easy, while others say up-front that it takes work to achieve our goals. In general, the easy programs provide us with excuses for our problems. Most people have had some hardship and problems in their life. Regardless of the absolute magnitude of the problem, our problems are significant to every one of us. As we mentioned in an earlier portion of this chapter, self-improvement takes work.

To get control of our lives, we have to understand why we act the way we do and why other people act the way they do. If we can understand the behavior of others, we can live a more rewarding life. Unfortunately, no one teaches us why we or others feel and act the way we do. Most of us are not taught—by either parents or formal education—some of the most basic but also the most important things in life, such as how to manage our emotions and resolve conflict.

People react to us based on the way we expect them to treat us, since we send out signals that in turn result in other people's behaviors toward us. The main idea is that we are not victims of the actions of others, but rather we create our own situations. However, this situation can sometimes be qualified by the fact that some really bad things do happen to some people, which they do not bring upon themselves. Furthermore, the world is random. But, in general, we should expect to be accountable for our actions to whatever extent possible. It may not be fair that certain things happen to people, but we must be accountable for how we react to these things. We should also add that thoughts are part of our behavior as well as outward actions. In fact, thoughts can be just as important as actions.

People repeat behaviors that work for them. This is a result of shaped behavior or conditioning. If we do something and we get positive feedback (payoff), we will continue. On the other hand, if we get negative feedback, we will stop. The key factor in making this principle work for us is to identify all the feedbacks we get.

People have to acknowledge what their life situations are today. People lie to themselves in two ways: by misrepresentation and by omission. Both are rooted in denial. Denial is a defense mechanism. Basically, you can't change what you do not acknowledge. On the other hand, many people cannot handle the truth, because accepting the truth may itself be painful. Furthermore, recognizing the truth may require change, and change is also painful.

People get in a rut, but they could live a more rewarding and stimulating life with some change. Life involves new things and risk. You have to buy a ticket to win the lottery.

Ironically, many self-help programs ask you to get out of a rut, but they also give you some form of system to follow. Thus, following a system in and of itself must not all be bad. If we live in a random world, we expose ourselves to risk, but the outcome could be good rather than bad. We all have setbacks, but if we are willing to suffer a few setbacks, we will most likely realize rewards as well. The net result will be better than if we had done nothing at all.

Meaning is a personal thing. Perceptions of reality vary with the individual, and often two people will not see the same situation in the same way. However, whatever meaning or value a particular circumstance has for you will be the meaning that you give it. We receive information through the senses, but the meaning that we give to those sensations is our perception of the situation. People interpret information differently. We all view the world through filters; some come from our personality and attitudes and some come from other factors: gender, ethnic background, religion, and age. These filters (fixed values) influence our interpretation of events. If we understand our filters, we will understand our interpretation of events. The bad thing about filters is that they can place limitations on our interpretation that we may not even be aware of. Some of these filters are as simple as "it is selfish to spend so much time thinking about yourself."

Life involves problems and challenges. We should expect this. If we anticipate problems, they will be easier to deal with as they occur. The key idea here is to do something about the problem rather than just endure it.

Through our behavior we let people know how we want them to treat us. If we respond to people in a way that they want, they will continue, but if our response has negative results for us, we have to determine how we can change our behavior. In cases of domestic violence, the abuse usually increases gradually as one of the partners, usually the wife, endures the abuse because leaving could result in financial insecurity or abuse of children. While on partner stays for reasons that are not meant to encourage the negative behavior the negative behavior is encouraged none the less

The most important step in getting what we want from life is saying specifically what we want. You seldom ever get more than you ask for. Therefore, it is important to set goals high enough so that you will be satisfied when you achieve them, but not so high that you will be frustrated if you do not achieve every one of them. You also have to be specific. You can't just say that you want to be happy. When we look at our lives, we should look at several aspects such as personal (including spiritual), work, relationships, and family.

The Road Less Traveled

One of the most widely read self-help books of all time is *The Road Less Traveled* by M. Scott Peck.

The principles and practices discussed in this hugely popular book are not fundamentally different from many of the ideas discussed in books of this genre. Peck notes that life is difficult; there are no easy answers; suffering can lead to growth; and personal growth takes work, self-discipline, and time.

If we have the capacity to be truly honest with ourselves, much of what we read in Peck's book would allow us to avoid the expense of professional guidance. However, as this book and others show through case histories, true self-knowledge is a long and difficult process.

One theme in Peck's book that gets more attention than some other self-help books is the important role that parents play in the development of their children. Too often, parents fail to provide the guidance or support that is required for proper development. This is not a conscious action. Rather it appears that we are not properly prepared ourselves by our parents.

Peck also looks at the role that religion plays in personal growth. Starting from the viewpoint that we all must have a world view, or some basic context in which we view our life in the universe, Peck suggests that everyone has a religion if we use this as a definition of religion. Peck believes that not enough attention is given to this world view, because it is the foundation upon which we build personal growth and it impacts our behavior. Although the world view provided by religion varies widely around the globe, we do not give enough attention to these differences in our dealings with other people.

Like many psychiatrists, Peck believes that the authoritative teaching of many Christian churches and the concepts of guilt associated with sin provide an unending source of new patients for the practice of psychotherapy. However, rather than taking sides, Peck looks at the good things that both science and religion can contribute to our personal growth and the bad things that should be avoided in both science and religion.

Peck and others have pointed out that if something cannot be studied by science, scientists say it should be avoided. In other words, if something is too complex to be studied, it should remain a matter of faith. But if we look at the increase in scientific knowledge over time, science is able to explain more and more things, even things that appeared paradoxical only a few years ago.

Nature involves various degrees of complexity. As we develop more sophisticated tools, we are able to understand and explain more and more complex phenomena. What was once science fiction becomes true science, and what was faith now becomes science.

In the recent past, both science and religion have engaged in their own forms of tunnel vision. Peck takes the optimistic approach that some day there may be a merging of these two approaches to reality. I do not take the position that the inexplicable is the result of divine inspiration, rather I think it is just a matter of time before we figure out what we still do not understand.

Some people may call it the religion of nature, but I believe that we are just starting to learn about both the complexity and the organization in the universe. This combination of complexity and organization is awe-inspiring. Some people may call it a religious experience. Rather than questioning religion or science, I believe we should take a look one step deeper and examine what this experience is that we call spirituality and what it means to our efforts to find a meaning for our existence.

The conflict between science and religion is presented to us in many ways. In the beginning of his book, Peck points out that he makes no distinction between mind and spirit. The mind-body or spirit-body controversy has been with us since ancient times. Hippocrates declared in the fifth century B.C. that mental illness was the result of the dysfunction of the brain, which could best be treated with medicine. Meanwhile, Plato argued that mental problems could be solved through reasoned discourse. In the Middle

Ages, mental illness was believed to be a religious problem, in which the afflicted person was possessed by some evil spirit.

In modern times, the profession of psychiatry continues to be split in how professionals should treat the ills of the mind: through talk therapy or the use of drugs. Most recently, the use of drugs has become the favored form of treatment, which is largely due to the significant progress that has been made in psychopharmacology. The popularity of these drugs has been enhanced by the publication of books such as Peter D. Kramer's *Listening to Prozac*, which documents the significant impact that drug has had on patients.

Depression is one of the most widely treated mental ailments today and one of the most controversial in terms of treatment. Some scientists believe that some people are genetically predisposed to depression, especially people with mild forms of the problem. Other scientists believe that depression is a symptom or byproduct of the stress that we endure as a result of our lifestyle. Regardless of the ultimate cause of depression, there is less dispute that depression contributes to physical ailments such as heart disease.

All of the concepts introduced in this book come down on the talking or philosophical side of mental health. I believe that it is important to touch briefly on some of the issues in mental health. Any discussion of the meaning of life should include some discussion of depression, since one of the symptoms of depression is a feeling that life has no purpose. People who suffer from depression lose interest in all things that give them enjoyment. Depression goes beyond the dissatisfaction that people feel when they confuse pleasure with happiness. People who confuse pleasure with happiness still seek pleasure and have enjoyment, even if is only fleeting. Whereas people who are depressed have lost all desire even to seek pleasure.

Cynics point to the wide use of drugs, both legal and illegal, as a sign that our society today has lost its sense of values and suffers from a lack of purpose. On the other hand, Karl Marx once said, "Religion is the opium of the masses." The only thing that can be said with certainty is that we live with suffering. I personally believe that we should use all our knowledge to reduce suffering whenever possible. I believe that modern science, both talk therapy and medicine, can aid in the search for meaning in our lives. While we may never get the meaning of life from a pill, we should not ignore modern psychopharmacology as a possible aid in the quest. (Note, however, that I am not endorsing the use of so-called conscious altering drugs.)

In *Road Less Traveled,* Peck admits that he believes that some things cannot be explained and are in fact miraculous. He asks the reader to contemplate these things rather than ignoring them just because they cannot be explained by science. On the other hand, he cautions us to be wary when we explore the inexplicable—charlatans are always eager to exploit people willing to explore the unknown.

Some of the miracles that Peck examines in detail include health, the unconscious, serendipity, and evolution. I believe that while these things cannot be explained today, they will be explained at some time in the future. In fact, since Peck wrote his book more than twenty years ago, we are a little closer to explaining some of these things today. In the overall scheme of the universe, twenty years is an infinitely short period of time.

In spite of the fact that our bodies are so complex and we live in such a hostile environment where infectious disease and other events could easily disturb the natural balance in the body, the large majority of people live healthy lives. Some people attribute our health to the survival instinct, but what is the survival instinct, and why does it exist?

Our unconscious through dreams is able to provide us with analogies or stories that tell us what is wrong with our lives. Because dreams are so good at getting at the root of people's problems, therapists spend a large portion of their work on dreams. The process of developing insight that seems to suddenly spring from idle thoughts provides new understanding of ourselves or the world around us. The important thing about dreams and insights is that they come without our willing them to come to us. Beyond the unconscious, the working of the human mind is so complex that its operation is a miracle; in fact, that the mind can comprehend itself is a mystery.

Serendipity or luck, as some people might call it, is the occurrence of events that could be much less probable than would be expected by pure random events. Everyone has some experience of this phenomenon in their lives, and some people seem to experience more luck than others or at the very least they are more aware of the fact that it is occurring.

In the process of writing this book, I noticed on several occasions that I happened upon a particular book or article that at first glance would not seem related to what I was working on, but indeed enhanced my research. Sometimes after reading a section of the book or the article, I was better able to clarify a point in my own writing. In other cases, the information I had happened upon so serendipitously led me to take a different tack, a new direction that seemed a departure from the main thread of the thought but which eventually took me more directly to the end point toward which I was headed.

In the course of writing this book, I have developed my own theory of serendipity. I have reviewed the thoughts of people who work or practice in a wide range of disciplines. As a result of the exponential expansion in our knowledge base, it has been necessary for people to study an ever-narrowing portion of the universe. I have found that people have independently developed theories (and actually proved them through scientific experiment) that at first seem to be completely unrelated, but on further examination reveal the same basic universal law at work.

Earlier in this book I reviewed the principles of self-organizing complex systems. Study has shown that these principles apply equally as well to microbiological systems or to national economies. Thus, I believe that when we observe events that do not appear to be connected in a causal relationship but also do not appear to be occurring at random, we are observing some relationship that is based on reality—however, it is occurring at some other level of reality than the one we think we are observing. I also believe that our reductionistic approach to science tends to encourage us to look for this relationship at some more basic level, when in fact the relationship may be occurring at some higher level of organization.

With regard to evolution, Peck correctly observes that the second law of thermodynamics, which states that the world tends to wind down or move toward a state of more disorganization, makes it appear that evolution of life should not take place. Thus evolution must be a miracle. By applying the laws of self-organizing complex systems, it is possible to explain how evolution could take place. It is also important to take into account our frame of reference. While the time that it has taken for life to evolve from elemental particles to the human species, approximately three billion years, may seem a long time, it is not long in terms of the evolution of the universe. If we view the universe from a sufficiently long time frame, we will see that evolution is a temporary phenomenon.

This same law also has led astrophysicists to question how it is even possible for the universe with its many billions of galaxies to even exist. If the universe started as a big bang, it should have either continued to expand uniformly or just quickly collapsed on to itself. However, science has shown that slight irregularities in the first split second of the universe could make all the structures we now observe in the universe possible.

Peck finishes his book with the concept of evolving to a higher level of consciousness, which he equated with moving closer to God. He looks at original sin and evil with a nonreligious spirituality, which tells us to accept the grace of God. Grace will allow us to help our fellow humans through love to our own higher level of awareness and thus push evolution of the entire human species forward.

In the most basic sense, Peck has developed the final thoughts in his book on a personal level that allows him to relate through his years of experience to develop a meaning of life that applies to himself and the human race. In the final analysis, Peck, who has helped millions of other people with spiritual growth, shows us that even his growth is a personal process.

Perception and Reality

One of the most important things that self-help books help us do, is see reality rather than what we think we see. A therapist could say that a patient is cured, or at least on the road to recovery, after they have confronted a painful experience, or recognized a behavior that resulting in conflict with others or emotional pain. Unlike the cause of physical pain, the true cause of emotional pain is usually not obvious to the person suffering.

Self help books and the social sciences apply scientific methods to study human behavior. In other words, they attempt to make a cause and effect connection between experience and subsequent behaviors or emotional states. People may not consider Eastern religions as science, but these religions do apply what I consider logic to the causes of our behavior. In any case, Eastern religion provides a perspective on our behavior, which would describe as allowing us to see our true experience rather then what we think we experience.

On of the primary objectives of the first part of this book is to contrast some common misconceptions about nature with what we have learned from science about reality. Once people have a better understanding of nature, they can use this new perspective of reality to either modify their behavior or at least enjoy a greater sense of wellbeing, through this understanding. The next and last chapter presents a prescription for getting the most from the experience of life, which is based on our diagnosis of the human condition at the start of the twenty first century.

Chapter Eleven

Life Is a Symphony

"No one imagines that a symphony is supposed to improve in quality as it goes along, or that the whole object of playing it is to reach the finale. The point of music is discovered in every moment of playing and listening to it. It is the same I feel with the greater part of our lives, and if we are unduly absorbed in improving them, we may forget altogether to live them."
— Alan Watts

A Journey Has an Itinerary

Life is a journey, not a destination. Don't get so focused on your goals in life that you forget to enjoy life as it passes by. In the worst cases, some people actually significantly degrade the quality of their life in an effort to achieve goals they never reach, or the goals prove to be less than expected once achieved.

One of the dangers of living in a country where any man or woman can achieve great things is that too many people will actually try to achieve great things. Western society is a goal-oriented society, which has its advantages and disadvantages. We can attribute most of the improvement in our quality of life to the fact that we believe that we can use what we have learned about nature to improve our quality of life. Unfortunately, the laws of economics in a society that focuses on technology require that we continue to consume more and more goods and services. The company that is best at selling the most goods and services at a profit and can promise to sell even more next year is the company that has the highest stock price.

This is nothing new. During the sixties, a portion of the youth in the United States attempted to reject the mass consumption culture and live in communes or some other form of alternative living style. Most of these alternative lifestylers eventually found themselves seduced back into the mainstream by their most basic biological drive—the desire to have children and raise their children the way their parents had raised them. Many people will attribute this return to the mainstream as a return to maturity, and write off any attempt at an alternative lifestyle as a period of rebellion.

Today we see several trends within mainstream society, which I believe are also an attempt to counter the impact of the mass consumption society. One of these trends is manifested in the popularity of books on simplicity. Another trend is the return to religious fundamentalism. The trend toward simplicity says that less is more. While ideas may vary slightly from group to group, fundamentalists place an emphasis on love of God and family either in opposition to material consumption or as a means of finding balance in life.

Webster's Dictionary defines journey as "a traveling from one place to another, usually taking a rather long time; trip: *a six-day journey across the desert.*" Since this definition includes "one place to another" a destination is a part of a journey. In fact, *Webster's* defines wander as "to ramble without a definite purpose or objective; roam rove or stray; *to wander over the earth.*" Thus, when one says that "life is a journey, not a destination," one is simply trying to shift the emphasis away from the goal toward the process of getting to the goal. Without a goal, one would simply be wandering.

In the previous chapter, we talked about many ways that we can plan our life. Planning our life will make it more rewarding and meaningful. Every business trip and vacation should have an itinerary. The itinerary for a vacation may have fewer details than a business trip, but, in any case, working with an itinerary enhances the value of the experience.

In this book, we have examined how science has uncovered some of the truths of nature and how these laws appear to be consistent through the universe that we can observe. Some scientists have speculated that one explanation for some of the things that we cannot explain is that there are some things in nature that we by our very nature cannot observe. For example, there may be parallel universes, or substantial amounts of antimatter may exist in the universe—and the mere act of observing these things would, in the process, cause us or them to cease to exist. Thus, we must reconcile ourselves to the fact there must be some things that we cannot know. In the chapter on higher meaning, we described how it may be possible to have design without intelligence. Without the information that we have about evolution that appears to support such a process, all logical people would say this is impossible. At the very least, it raises the possibility of limitations to logical thought for humans.

Based on what we have been able to learn so far, it does not appear that there is any ultimate purpose of nature as we know it on earth other than the continuation of life. On this one thing, the continuation of life, we have learned that nature has gone through a long and complex process to arrive at the point where we exist as a species on the planet earth. We also know that nature has developed in such a way that it has provided for life to continue even under extreme changes in environmental conditions. On the other hand, it cannot be said that the forces of nature are singularly focused on the preservation of the human species.

While some may become discouraged or even despair that the study of science cannot show any purpose of life beyond mere existence, most scientists find the ability to see the workings of nature—how nature fits together, how the laws of nature are interconnected, and how the laws that govern the small things are the same laws that govern things in the scale of the universe—to be comparable to viewing the greatest works of art or listening to the best symphonies. Scientists find this satisfying. Their enjoyment is compounded by the awareness that the human species is the first species that has had sufficient self-awareness and intelligence to be able to wonder at the beauty of nature.

Social Evolution and Keeping Up with Technology

Evolution brought us to the point where we as a species can enjoy the flow of nature and our journey through life. Social evolution, which has greatly accelerated our dominance over nature and our ability to control it for our benefit, also changed our orientation to focus more on goals and objectives.

Physical evolution has not changed the human species much since it was able to develop our large brain size and self-awareness. If it were possible to take a baby born to a couple of Homo sapiens forty thousand years ago and raise it in a modern-day family environment, the child could develop no differently than a child born to a modern-day couple.

Through social evolution, our species has learned that if we work together, the combined efforts of the group provide more benefit than working alone. Through our intelligence, we have been able to conclude that some things follow others on a fairly regular basis—in other words, we have been able to anticipate outcomes. For the social organization to work over time, humans had to develop a social commitment. The ability to anticipate was an essential precondition for the social contract.

The individuals that worked with the group shared in the benefits of the group. Living in the group also required social controls. Behavior that supported the goals of the group and contributed to harmonious interaction of the group was rewarded by the group, and behavior that did not contribute to the group was punished. We would not be too far off by assuming that the people who worked at odds to the group were banished from the group. The first groups, like tribes today, were no more than extended families.

As social organization became more complex, so did social contracts. Slowly we evolved tools that could be used to assist with our social contracts. For example, the first trade among groups was through a barter system in which some goods were directly exchanged for other goods. After awhile, coins made from precious metals were used to assist the trade process. Coins were replaced by paper money, and today paper currency is being replaced by plastic credit cards and electronic accounting. In some cases, wealth has become no more than so many bits of information stored in a computer.

While social evolution may have accelerated the development of the human species by allowing technological innovation, social evolution has not moved fast enough to allow our species to deal with all the impacts of our technological innovation. The pessimists today will say that we are on the verge of blowing ourselves up with nuclear weapons, or changing our environment so radically that we will not be able to survive as a species. But, these pessimists forget that in a time period that is but a blink of any eye in terms of the evolution of our species, we have eliminated slavery, mistreatment of children, and inequality of the sexes. There is no question that we are living in a dangerous period for our species, until social evolution can mature enough to be on a par with our technological capabilities.

Regardless of how we treated one another, most humans lived on the edge of existence until just a few generations ago. The Greeks and the Romans enjoyed periods of cultural advancement and were able to provide a quality of life that would be considered bearable by today's standards. But these advances were enjoyed by only a small segment of the population and these benefits were at the expense of other groups that had been conquered in war.

Today through the use of technology, we have been able to provide relative abundance to the common man. More importantly, through the use of technology, we have been able to provide culture and aesthetic enjoyment to the common man that were impossible for even a king two hundred years ago.

Until recently, the meaning of life for the average man was simple—provide for his basic needs and the basic needs of his family. The same social groups that provided for his basic physical needs also provided for his social and esthetic pleasures. On the other hand, it did not take much knowledge and training to become a productive member of that social group. Today, it takes much more training for an individual to be a productive member of society. As the peoples of the world merge into a global community, we will be required to bring all people up to some basic level of productive capability. It has taken 225 years in the United States to provide some measure of equality. As we work to complete the job in our own country, we also must work toward and support a similar condition among the rest of the people in the world. Training people to be productive members of society benefits society as a whole. What has worked for the United States has worked for other countries, and it can work for the poor nations of the world, which have not as yet managed to reach our level of productive capacity.

The Pursuit of Happiness (Property)

The Declaration of Independence is one of our primary social contracts in the United States. The second paragraph of the Declaration of Independence starts off with the words that nearly every American is familiar with: "We hold these truths to be self-evident, that all men are created equal, that they are endowed by their Creator with certain unalienable Rights, that among these are Life, Liberty, and the pursuit of Happiness."

Thomas Jefferson borrowed the words of John Locke to enumerate human rights in the Declaration of Independence. However, he made some slight but important changes. Locke wrote "life, liberty . . . property." Jefferson replaced "property" with "the pursuit of happiness." Although we may not be clear about the meaning of the pursuit of happiness, it appears that there was some intent to connect the pursuit of property with happiness. One should also remember that, while some people came to America for religious freedom, the colonies were set up as an economic endeavor. It should be pointed out that Jefferson and his contemporaries believed that happiness could also be attained by improving the lives of others, and that the desire to help others was an ingrained part of human nature.

To participate in today's society and share in its abundance, people must spend more time learning to become a productive part of the society and the average man must be able to interact with a wider range of social groups. Today people expect to spend their whole life learning new things. Machines and computers do many of the things that the average man did a few years ago. These changes have taken place so rapidly that even social evolution has not been able to move fast enough to accommodate these changes.

The same social forces that moved us when we lived in a time of scarcity are still used to move us in a time of abundance. The majority of people in the United States already own their home. So the trend is toward bigger homes with more and bigger appliances. Everyone wears designer clothes. While productivity goes up every year, a larger percentage of the population is in the workforce now than at any time in our country's history. It's not hard to see how we have come to be a society in which the meaning of life is defined in terms of more and more goods and services. It's also not difficult to see how the average person feels like he or she has to run faster just to stay in place—in terms of standard of living. Keeping up with the Joneses has never been more difficult.

Now that we live in a time of relative abundance, it is time to focus on life as a journey, rather than a destination. When we become focused on journey instead of destination, we can relax and forget about keeping up the Joneses.

In difficult times when life is basically a struggle, it is not hard to understand that society would tell people that we cannot be free from the stress of day-to-day living and that we can only expect to get our reward in a later life, where all things are perfect. Personally, I cannot envision a world in which all things are perfect and I am also blissfully happy. I believe that it is part of human nature to grow bored with things that are too perfect. People enjoy a challenge as long as the challenge is kept to manageable levels. Jack Welch, the former chairman of GE, said in his memoirs that the major part of his business success came from his ability to properly challenge his managers. I will also concede that for some executives the pursuit of wealth is the ultimate game and the million-dollar compensation packages are just a way of keeping score. With these executives, wealth, the destination, is much less important than the journey, the power game. Unfortunately, playing this game requires corporate executives to make sacrifices in other areas of their personal lives, because goals, such as a senior executive title, become goals that take away from the value of the journey.

Through technology we have already been able to remove much of the stress, which also involves excitement, in our lives. Children and even adults are using electronic games and sports as a means of putting some of the excitement back in their lives. The increase in lotteries and gambling casinos illustrates our desire to experience the excitement of leaving some of our fate to the laws of chance. I do not believe that the absolute dollar amount that anyone could win is the prime motivator for most people who gamble. Rich people, who as a practical matter have no need for more wealth, still go to casinos.

About the only satisfaction that we get from material things is the ego satisfaction that we get from the fact that we were able to acquire the goods. However, as everyone of our neighbors and friends are also able to acquire these goods, the satisfaction is fleeting. In the case where we suffer a temporary financial setback, many times through no fault of our own, we suffer because we are now behind in the materialism race.

Automation and the Obsolescence of Work

As an aside, some people have suggested that a not insignificant segment of the population will eventually have to learn how to live the way English gentlemen lived in the nineteenth century. Basically, in the nineteenth century, you could not be truly considered a gentlemen if you had to work for a living.

Through technology, we have significantly increased the productivity of the average farmer. As the farmers became more productive, they could move into the manufacturing sector. This may not have been seen as an improvement at the time, due to the poor working conditions in factories and exploitation, but it resulted in a movement of the labor force nonetheless. As we introduced technology into manufacturing, we have either moved workers into the service sector or moved jobs overseas where lower wages can compete with the more efficient manufacturing in the U.S. In recent years, through the use of information technology, we have been able to move people out of the service and administrative sector. While we have been able to introduce some technology into some service sectors such as fast food, a significant portion of the workforce continues to be employed in the jobs that do not lend themselves to automation.

Until now, foreign markets have been able to allow our economy to provide for these low-wage workers, either through sending the jobs overseas in manufacturing or through employing emigrants in this country in these low-paying jobs. Although wages are low by U.S. standards, there seems to be a sufficient supply of people who are still willing to do these jobs, because the wages are high compared to those in their home market.

The world is making progress in raising living standards. Eventually, and quite soon, in terms of the time frames that we have been talking about in this book, there will not be enough work for everyone. At one time, automation eliminated unskilled jobs, but for the last twenty years, computers have started to eliminate jobs that required significant skill or market knowledge. In spite of the dot-com meltdown, the Internet has already done an excellent job of eliminating the jobs of many people who acted as middlemen in the provision of goods and services.

Progress and Current Times

As a result of living in a time of rapid technological change, it is easy to believe we are living in the most exciting times. However, I believe that any generation, if not all generations, can feel that they also lived in exciting times. The process of establishing a new country, developing the American western frontier, and the industrial revolution must have also been exciting times. Being human will always be a process of being, while also becoming something new.

Earlier in this book, we looked at our life span in terms of the development of life. Each of our lives is relatively a blink of an eye, even to the development of our species. If evolution had some objective other than just adaptation to change, only the last generation that lived when evolution reached its goal would benefit from all this progress. I do not believe that if there was some design for life or our species that just the last generation or the last species was the intended objective of that design.

When we talk about human progress, we also have to consider personal progress, death, and the concept of reward after death. If we work toward some destination that is at the other side of this life, that goal should not take away from our enjoyment of the journey through this life. The concept of life after death is as much a way of dealing with the grief of the loss of loved ones as it is with dealing with our own mortality. Anything that we should be expected to do in this life that would result in a reward after death should also contribute to enjoyment of our journey through this life.

Don't Wait to Enjoy Life

To comprehend the idea of spending life as a journey rather than focusing on a destination, think about how you would spend two one-week periods. One of the weeks is seven days spent on a vacation at a resort, and the other is a week spent on the job. I think the week at the resort would be spent day by day trying to savor the beauty of a new location and the freedom from responsibility. The week on the job would most likely be spent anticipating the end of the week and the stress of solving problems big and small. Both week-long experiences present us with new situations, but our approach to these situations has a major impact on our quality of experience during the week.

While we are on vacation, we spend our time hoping the time will last as long as possible, while the week spent on the job is spent waiting for the time to pass as quickly as possible so that we can get to the weekend when we can do something we would rather be doing. Some people might say that the things that you have to do on the job are less pleasant tasks than you do while on vacation. However, there are many things that people do with their time that for some people is a vocation and for other people is an avocation. In other words, some look forward to doing for free what another person gets paid to do. Thus, it is only our approach to the activity that gives it meaning.

I am not suggesting that we should expect every week on the job to be like a vacation. However, I am suggesting that approaching life as a journey to be enjoyed as much is possible is far superior to a series of weeks and months that are wished away in an attempt to reach a goal that may never be achieved. I believe that this approach to life is especially important when we realize that we often have little control over random events or other factors that we cannot control. As important as it is to realize that we may not have control over events to come in the future, it is as important to realize that we may have to live with the impact of events that have happened to us in the past.

No matter how routine our lives may be, every day is a truly unique experience. Eastern philosophy teaches that we get to live our lives many times over. The idea is that we get to come back to live our lives over so that we live life better the next time around. However, the random aspect of nature says that we never get a chance to do the same thing over again under exactly the same circumstances. In reality, we live in the eternal present.

State Destinations in General Terms

If destination is an integral part of a journey, we must also discuss goals and objectives. The best objectives can be stated as general direction rather than specific targets. For example, well-stated objectives might include "I would like to live a life that gives me the greatest opportunity to experience new things" or "Security is important to me so I will seek to be in situations that present the best chance for guaranteed employment." We may want to set an objective to work in a particular field of study, but don't express that objective in terms of achieving a certain level of accomplishment.

Regardless of how specific our goals, our aim should be to take in the whole experience as we work toward our goals. In some cases, specific goals can focus our efforts. In other cases, we may need the focus to put forth the effort required. However, no matter how badly we want to achieve something, failure should not take away from the experience of working toward the goal.

Our Legacy

We do not know if we were given dominion over the earth by a higher power or if we have been fortunate to be the current proprietors of life on earth through some glorious accident. We seem fairly certain that if we are able to learn anything through scientific inquiry that the human species is the result of a process of increased complexification and sophistication that has taken billions of years.

We know that each generation holds the keys for only the briefest instant before we pass them along to the next generation. Most people think of their legacy as a few possessions and accomplishments they themselves have been able accumulate in their time on this planet; they never think about the precious gift of life that they were given. One by one, nature encourages us to pass this gift along and to nurture it until those that we give life to are able to pass it along to the next generation. Whether nature or God gave us the added gift of intelligence and consciousness, we have been able to use this gift to convert our daily lives from a struggle to an awe-inspiring experience.

It is difficult to comprehend what a spiritual life would be like after death if it had no physical component, since our only frame of reference is the physical world. If I think back to some of the high points in my life—watching my children graduate from college, sharing the beauty of a sunset with a loved one, listening to Mozart, bringing a new product to market, helping a person in need, understanding the true meaning of a new idea—they all had some physical reference. This may not be all there is to life, but this is all I know. Some religions may look forward to the end of life on earth; I don't.

Science tells us life on earth will end from natural causes eventually. The end could come at any time, but the chances are that it will be millions or billions of years in the future. A time of one million years is far beyond human comprehension on a personal basis for any generation while it is alive, other than as a legacy for generations to follow. The human species is the first species to have sufficient control over nature to end the life of our species and possibly all life on earth. And the generation now alive is the first of our species to actually realize this potential.

Past generations had little choice but to pass life on to the next generation regardless of whether they believed that life is a journey or a destination. The people who today believe that life is a destination have the potential to end the journey for the human species. For all the people alive today and for all future generations, we must make everyone see that life is a journey and not a destination.

Bibliography

Adams, Scott. 2000. *Random Acts of Management*. Kansas City: Andrews McNeal Publishing

Angier, Natalie. Confessions of a Lonely Atheist. The New York Times Magazine January 14, 2001

Baumeister, Roy F. 1991. Meanings of Life. New York: The Gilford Press

Bentley, Peter J. 2002. *How Nature is Transforming Our Technology and Our Lives.* New York: Simon & Schuster

Campbell, Joseph. with Bill Moyers Betty Sue Flowers 1988. *The Power of Myth*. New York: Doubleday

Croswell, Ken 2001. *The Universe at Midnight*. New York: Simon & Schuster

Csikszentmihalyi, Mihaly. 1990. Flow The Psychology of Optimal Experience. New York: Harper & Row Publishers

His Holiness the Dali Lama and Howard C. Cutler M.D. 1998. *The Art of Happiness, A Handbook for Living*. New York: Penguin Putman Inc.

Dennett, Danial C. 1995. *Darwin's Dangerous Idea*. New York: Simon & Schuster

Delbanco, Andrew. *Are you Happy Yet?*. The New York Times Magazine. May 7, 2000

Ehrenreich, Barbara. 2001. *Nickel and Dimed, On Not Getting By in America.* New York: Henry Holt and Company

Headlam, Bruce. *Nothing Personal* The New York Times Magazine February 17, 2002

Jeans, Sir James. 1943. *Physics and Philosophy*. Cambridge University Press and the Macmillan Company

Kaku, Michio. 1994. *Hyperspace A Scientific Odessy Through Parallel Universes, Time Warps, and the 10th Dimension.* New York Oxford: The Oxford University Press

Kayzer, Wim with Olive Sacks, Stephen Jay Gould, Danial C. Bennett, Freeman Dyson, Rupert Sheldrake, Stephen Toulmin 1997. Companion to the Public Television Series, *A Glorious Accident*. New York: W H Freeman and Company

Lewis, Thomas, Fari Amini and Richard Lannon. 2000. *The General Theory of Love* New York: Random House

Lumsden, Charles J. *Sociobilogy, God, and Understanding.* Zygon, March 1989, Vol 24 Number 1

Lyall, Sarah. *For Europeans, Love, Yes; Marriage Maybe*. The New York Times March 24, 2002

McGraw, Phillip C. 1999. *Life Strategies, Doing What Works Doing What Matters*. New York: Hyperion

McGraw, Phillip C. 2001. *Self Matters, Creating Your Life from the Inside Out*. New York: Simon Schuster Source

Nolte, David D. 2001. *Mind at Light Speed, A New Kind of Intelligence*. New Yprk: Simon & Schuster

Novello, Joseph R. 2000. The Myth of More. Mahwah, NJ: Paulist Press

Peck, M. Scott. 1978. *The Road Less Traveled A New Psychology of Love Traditional Values and Spiritual Growth*. New York: Simon & Schuster

Sagan, Carl. 1977. The Dragons of Eden. New York: Ballentine Books

Smoot, George and Keay Davidson. 1993 *Wrinkles in Time*. New York: William Morrow and Company

Sobel, Dava. 1999. *Galileo's Daughter, A Historical Memoir of Science Faith and Love*. New York: Penguin Books

Suskind, Ron. *A Plunge into the Present.* The New York Times Magazine. December 2, 2001

Thoreau, Henry David. 1854. *Walden* or *Life in the Woods*. Boston Ticknor and Fields

Waldrop, Mitchell M. 1992. *Complexity*. The Emerging Science at the Edge of Order and Chaos New York: Simon & Schuster

Wilson, James Q. 2002. *How Our Culture Has Weakened Famalies.* New York: Harper Collins

Weaver, Henry Grady. 1953. *The Mainspring of Human Progress.* Irvington-on-Hudson, NY: The Foundation for Economic Education, Inc.

Wolfe, Alan. *The Pursuit of Autonomy*. The New York Times Magazine. May 7, 2000

Wolfe, Alan. *The Final Freedom*. The New York Times Magazine March 18, 2001

Index